Fractal Analysis - Applications and Updates

Edited by Sid-Ali Ouadfeul

Published in London, United Kingdom

Fractal Analysis - Applications and Updates
http://dx.doi.org/10.5772/intechopen.1000341
Edited by Sid-Ali Ouadfeul

Contributors
Amela Zekovic, Andrey Chekuryaev, Arturo I. Quintanar, Ashiq Hussain Bhat, Carlos E. Puente, Carolyn Salafia, Donatella Bongiorno, Emily S. Barrett, Emőke Imre, Gamze Çoban, Ismail A. Mageed, James Wanliss, Jessica Brunner, Maxim Sychov, Pratipaul Persaud, Richard K. Miller, Ruchit Shah, Sergey Mjakin, Sid-Ali Ouadfeul, Theresa Girardi, Thomas O'Connor, Victor M. Peñaranda-Vélez, Vijay Pal Singh, Óscar J. Mesa, Şerife Ebru Okuyucu

First published in London, United Kingdom, 2024 by IntechOpen
IntechOpen is the global imprint of INTECHOPEN LIMITED, registered in England and Wales, registration number: 11086078, 5 Princes Gate Court, London, SW7 2QJ, United Kingdom

British Library Cataloguing-in-Publication Data
A catalogue record for this book is available from the British Library

Additional hard and PDF copies can be obtained from orders@intechopen.com

Fractal Analysis - Applications and Updates
Edited by Sid-Ali Ouadfeul
p. cm.
Print ISBN 978-1-83769-613-0
Online ISBN 978-1-83769-612-3
eBook (PDF) ISBN 978-1-83769-614-7

For EU product safety concerns:
IN TECH d.o.o., Prolaz Marije Krucifikse Kozulić 3, 51000 Rijeka, Croatia, info@intechopen.com or visit our website at intechopen.com.

Meet the editor

Dr. Sid-Ali Ouadfeul is Professor of Geophysics in the Department of Geophysics, Geology, and Reservoir Engineering, Algerian Petroleum Institute-IAP Corporate University, Algeria. He received an engineering diploma from the University of Boumerdès, Algeria; a master's degree and Ph.D. in Geophysics from the University of Science and Technology-Houari Boumédiène, Algeria; and a Ph.D. in Physics from the University of Boumerdès, Algeria. He is a pioneer researcher in fractal analysis and its applications in physics. Dr. Ouadfeul has edited books and journal special issues and is the author of many publications and proceedings on fractal analysis and its applications in earth sciences.

Contents

Preface

In recent decades, fractal analysis has become a hot topic of research. This book presents some recent applications of fractal analysis in various areas of science. The book opens with an introductory chapter. Chapter 2 is an overview of the fractal methods in nonequilibrium systems, while Chapter 3 discusses derivation and integration on a fractal subset of the real line. Chapter 4 deals with generalized Z-entropy and fractal dimensions, while Chapter 5 deals with the fractal analysis of facades of traditional historic buildings using the box-counting method. Chapter 6 discusses the fractal characterization of the microstructure of materials with their properties based on the digital materials science concept. Chapter 7 deals with the relation between the finite, discrete, and fractal distributions and mean grain size distributions of granular matter, while Chapter 8 discusses placental growth and development analyzed through 2D and 3D fractals. Chapter 9 describes the multi-scaling of the space-time structure of rainfall fields. Finally, Chapter 10 addresses fractal and multifractal characterization of 3D video signals.

This book is a valuable source for researchers and students working on fractal analysis and its applications in the full spectrum of sciences.

Sid-Ali Ouadfeul
Algerian Petroleum Institute,
Sonatrach, Boumerdes, Algeria

Chapter 1

Introductory Chapter: Fractal Analysis – Applications and Updates

Sid-Ali Ouadfeul

1. Introduction

Numerous objects in nature have a fractal behavior; a few of the foremost common illustrations would include: (1) branches of trees (**Figure 1**), (2) animal bodies (see **Figure 2**), (3) snowflakes (**Figure 3**), (4) lightning and power (**Figure 4**), (5) plants and takes off (**Figure 5**), (6) geographic mountains (**Figure 6**), (7) river frameworks (**Figure 7**), (8) clouds (**Figure 8**), and (9) precious stones (**Figure 9**) [1]. Fractal analysis is the modeling of information of which fractality is the inborn property. The key idea of fractal goes back to Mandelbrot [2], who presented it in 1977 as a numerical representation of self-similar objects. The fractal analysis applies to physical frameworks that are recognized by a similitude of behavior over a multitude of scales or, within the most articulated cases, by self-similarity, where this similitude is protected through an interminability of scales. Domains of application of the fractal analysis are developing, from mathematical concepts in early 1970s to the full spectrum of sciences, and it has illustrated numerous areas of applications. There are five primary applications of fractal analysis [3]: (1) fractal cities, (2) fractal medicine, (3) image compression and resolution, (4) antennas, and (5) art. This book will provide some advances and the latest applications of fractal analysis in some branches

Figure 1.
Example of branches of a tree that has a fractal architecture [1].

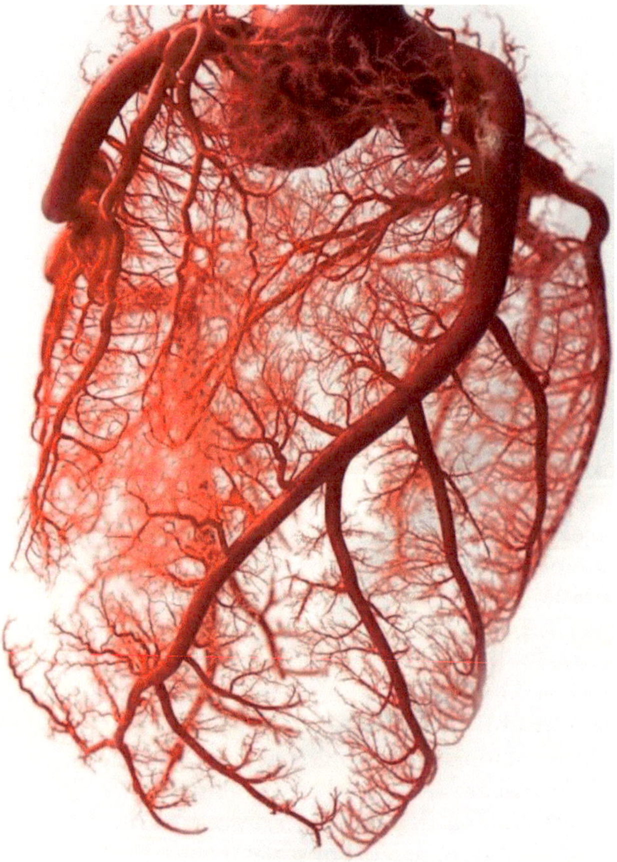

Figure 2.
Fractal behavior of animal's heart branches [1].

Figure 3.
A picture showing snowflakes with its fractal architecture [4].

Figure 4.
Lightning storm [1].

Figure 5.
A picture showing the fractal behavior of a plant [1].

of science. In what follows, we provide a brief description of fractal analysis applications within the five branches of sciences.

2. Fractal cities

Cities are complex frameworks that carry on in a few ways like living beings. The rules of chaos hypothesis and fractals apply straightforwardly to the development of cities, and the ponder of urban designs permits us to take advantage of the tests of past societies to shape our future with as much mindfulness of the results of our activities

Figure 6.
A picture showing the fractal behavior of a series of mountains [1].

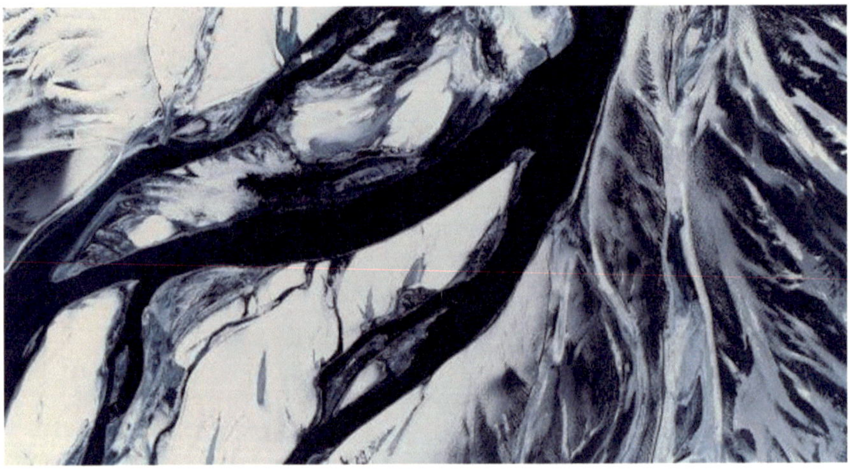

Figure 7.
A picture of a river with its fractal architecture [1].

Figure 8.
Clouds with its fractal behavior [1].

Figure 9.
A picture showing the fractal behavior of crystals [1].

as conceivable. The physical structure of a city provides the record of decades or centuries of human action. Cities that have a historical record of their urban approaches—basically zoning codes—allow us to think about the effect of distinctive sorts of rules on the advancement of cities. The fractal shapes of fractal cities come from millions of people choosing how they need to live. Essentially, it may be a city that has comparable structures at distinctive scales. A huge fractal city tends to develop that way by retaining parcels of smaller towns. So, the city gets to be a collection of towns, and the towns are collections of neighborhoods. The streets in a fractal city tend to go in apparently indiscriminate bearings, but truly, they serve as coordinate associations between different central center points in an energetic urban organize. An inclining street can be as much as ~70% (1/sqrt (2)) shorter than two streets in a rectangular framework, so there are efficiencies built into the geometry of the city.

3. Fractal medicine

Present-day medicine regularly includes looking at frameworks within the body to decide if something is breaking down. Since the body is full of fractals, we will utilize fractal math to evaluate, depict, analyze, and maybe, before long, assist in remedying illnesses. With advanced imaging hardware such as CT Scanners and MRI machines, specialists can access a tremendous sum of computerized information, almost an understanding. Making sense of all the information can be time-consuming and troublesome indeed for prepared specialists. Instructing computers to use scientific forms to tell the contrast between sound lungs and lungs enduring from emphysema guarantees to assist in making faster, more dependable analyses. The fractal measurement of the lung shows up to differ between sound and wiped-out lungs, possibly helping within the mechanized location of the illness. Cancer is another illness where fractal analysis may not only offer assistance analysis but may also offer assistance treatment. It is well known that cancerous tumors—unusual, fast development of cells—regularly have a characteristic development of modern blood vessels that frame a tangled mess rather than the flawless, efficient fractal organize of sound blood vessels. Not only can these failing vessels specifically hurt the tissue, but they can make it harder to treat the illness by anticipating drugs from entering the inward parts of tumors where the drugs are most required [5, 6].

4. Image compression and resolution

Fractal compression is a picture compression strategy that is still widely utilized nowadays. It is based on the discovery of the repeat of designs and tends to dispense with the redundancy of data within the picture.

This is often a damaging strategy since all of the introductory information is not found within the last picture. There are a few strategies (subdivision of triangles, Delaunay, etc.), but compression utilizing the Jacquin strategy is the finest known. The outline of the picture compression by the Jacquin Strategy is depicted as taken after:

- Fractal compression comprises, to begin with of all in carrying out two segmentations (moreover called tailings or dividing) on an image: a division of Source figures and a division of Goal figures.

- It is at that point a matter of finding for each Source figure what is the leading combine (source figure, goal figure) minimizing an error. This blunder is ordinarily calculated by subtracting the two figures. To carry out the subtraction operation, it is essential to convert the source figure to the measurements (and geometry) of the goal figure.

- Once all the sets have been found, the yield record at that point contains the different pairs, as well as the diverse changes carried out (revolution, diminishment of the cruel, etc.).

- Amid decompression, the picture is reproduced from these changes. Meeting is at that point ensured by the reality that there is a minimization of mistake (contrast) and an alteration of the pixels; on the other hand, the source figures are bigger than the goal figures. Fractal compression employs the same property to reproduce the picture [7].

5. Antennas

A fractal Antenna is a receiving wire employing a fractal plan called self-similar. The objective is to maximize the effective length or extend the fabric's edge (on the inner areas or the outside structure) that can get or transmit electromagnetic radiation on a given surface or add up to volume [8]. Such fractal antennas are moreover called filling bends or multi-level bends. Their key perspective is their reiteration of a design over a few scale sizes. For this reason, fractal antennas are exceptionally compact, multiband or broadband, and have broadcast and microwave communications applications. The reaction of a fractal antenna varies altogether from conventional

Antenna plans since it can work at numerous concurrent frequencies with great to fabulous execution. Ordinarily, standard antennas must be "trimmed" concurring to their recurrence of utilize, so standard antennas work well at that recurrence. In expansion, the fractal nature of the radio wire allows its measure to be decreased without having to utilize components such as inductors or capacitors. Fractals are created using the iterative strategy to unravel nonlinear conditions or polynomial conditions. The increment in computer control has allowed the creation of program permitting the calculation of three-dimensional pictures in computer-generated pictures, in this way advertising the capacities and impacts more often than not saved for classic three-dimensional modeling computer programs (lights, volumetric lights, profundity obscure, air, reflection/refraction of certain materials, surfaces, etc.) [9].

6. Art

Fractal art could be a set of algorithmic craftsmanship comprising pictures, liveliness, and music from fractal objects. Julia can be considered as symbol of fractal art. Fractal art can be drawn or painted by hand, or utilizing computers, which can calculate fractal capacities and create pictures from them. It is additionally the appearance of computers that allowed the advancement of this arts, since it requires a part of computing control.

Fractal picture generator programs ordinarily work in three steps: altering the parameters that oversee the era of the picture, performing the calculations, and, at last, applying the comes to arrange to produce an image. For liveliness, the operation will have to be rehashed for each image produced [10].

Author details

Sid-Ali Ouadfeul
Algerian Petroleum Institute, Sonatrach, Boumerdes, Algeria

*Address all correspondence to: ouadfeul77@gmail.com

References

[1] INTERNAL. What is a Fractal? Available from: https://iternal.us/what-is-a-fractal/#:~:text=Some%20of%20the%20most%20common%20examples%20of%20Fractals%20in%20nature,river%20systems%2C%20clouds%2C%20crystals

[2] Mandelbrot BB. Fractals. Form, Chance and Dimension. San Francisco: Freeman; 1977. 365 p

[3] MATHEMATICS. Top 5 Applications of Fractals. Canada: University of Waterloo; Available from: https://uwaterloo.ca/math/news/top-5-applications-fractals

[4] New York Times. Photos of Snowflakes Like You've Never Seen Them Before. New York, USA: The New York Times; Available from: https://www.nytimes.com/2021/03/10/science/snowflakes-photos-nathan-myhrvold.html

[5] Fractal Applications, Fractal Cities. Available from: http://fractalfoundation.org/OFC/OFC-12-3.html#:~:text=Rome%20is%20much%20more%20of,Image%20courtesy%20of%20Google%20Earth

[6] Fractal Applications, Fractal Medicine. Available from: http://fractalfoundation.org/OFC/OFC-12-4.html

[7] Saturn Cloud. Fractal-Based Image Compression Algorithm. 2023. Available from: https://saturncloud.io/blog/fractalbased-image-compression-algorithm/

[8] Abed AT, Abu-AlShaer MJ, Jawad AM. In: Al-Rizzo H, editor. Fractal Antennas for Wireless Communications, Modern Printed-Circuit Antennas. Rijeka, Croatia: Intechopen; 2020. DOI: 10.5772/intechopen.90332

[9] Fractal Antenna. Available from: https://www.antenna-theory.com/antennas/fractal.php

[10] Briggs J. Fractals: The Patterns of Chaos. London: Thames & Hudson; 1992. 192 p. ISBN: 978-0-500-27693-8

Chapter 2

Fractal Methods in Nonequilibrium Systems

James Wanliss

Abstract

The principles of complexity science can be applied to systems with natural hazards, and also human social and economic systems. Disruptive, extreme events result from emergent properties of complex, nonequilibrium systems that consist of interdependent components whose interactions result in a competition between organized, interaction-dominated behavior and irregular or stochastic, fluctuation-dominated behavior. The use of fractal analysis allows insights into the development of such extreme events, and provides input and constraints for models. The main goal of this paper is to share and expand the scope of some of the fractal methods commonly used in complex systems studies. A proper appreciation of the strengths and limitations of fractal methods can improve the assessment and analysis of risks associated with systems that exhibit extreme events.

Keywords: fractals, nonequilibrium systems, extreme events, R/S analysis, Higuchi, detrended fluctuation analysis

1. Introduction

Many physical and social systems demonstrate the existence of properties that can be described as complex, featuring behaviors with both regular and irregular components. Complex behavior has been shown to arise when a system has many elements whose links, internal competition for resources, and interdependencies are inherently difficult to analyze. Because of these complex, often nonlinear, interconnections studying one part of the system may not necessarily lead to a better understanding or ability to accurately forecast the future development of any part of the system.

Historically science has progressed through a reductionist approach whereby the system is diminished to its simplest components. Then equations and models are sought which focus on adapting the values of these system parameters to match the observed data. Although this has been an extremely fruitful avenue of research, there is increasing recognition that many systems feature a wide variety of phenomena which do not act in isolation. Rather they should be understood in terms of multiple interactions between different internal system elements, as well as interactions from outside the system.

Whatever the discipline, there is an emerging perception that in our cosmos empirical data frequently demonstrate certain universal patterns of behavior [1, 2].

Part of the difficulty is captured in the probability distributions of system parameters. In equilibrium thermodynamics and statistical mechanics, the internal system dynamics constantly tend towards one or more equilibrium states with simple probability distributions. An example of an equilibrium probability distribution is the unimodal Gaussian distribution function in which most system events are within one standard deviation of the mean. This is the case for the classic Gibbs-Boltzmann statistical mechanics and the Gaussian distribution is one example in which the Gibbs-Boltzmann entropy is maximized to ensure the equilibrium state. Systems whose probability distribution assume the Gaussian form rarely have events that fall outside three standard deviations from the mean. Events like this, which deviate significantly from the mean, can be called 'extreme' events [3]. But observations in social and physical dynamics demonstrate that many systems have internal processes or external influences for which physical equilibrium descriptions of macroscopic behavior do not apply; extreme events are much more common than would be anticipated in equilibrium systems. This is most notably observed in heavy-tailed distribution functions which demonstrates that the mean is not an adequate measure of the future behavior [4].

It is not easy to settle on a definition of an extreme event, but there are certain common features [5]. Extreme events occur suddenly and are usually unexpected. They exceed a threshold value close to the upper or lower range of historical data, for instance being defined as those that occur in the top or bottom 5% or so of measurements. Other ways scientists describe extreme events include how different they are from the average, how often they occur, or their likelihood of happening again. As mentioned above, they produce heavy-tailed distributions. Because the probability of appearance of an extreme event is extremely small, the waiting time for its appearance is often too long for all practical purposes; the event occurs on a scale large compared to the system size. Thus, the sudden appearance of an extreme event can be considered exogenous to the system [6]. An open system is by nature a nonequilibrium system since there are exogenous inputs to, and losses from, the system. Thus, establishing the formal Gibbs-Boltzmann entropy can be difficult or impossible.

In nature, the most frequently observed, fascinating, and intricate phenomena are typically collective stochastic systems. These are systems where many individual components interact, with processes characterized by randomness and dissipative energetics. They show long-range correlations, intermiitency, and combinations of features which violate physical equilibrium descriptions of macroscopic behavior. This has led to concepts in terms of nonequilibrium phase transitions [7–10], and nonextensive generalizations of Gibbs-Boltzmann entropy [11, 12]. The complexity of various physical and social systems is well recognized, arising mainly from the inherent nonlinearity and far from equilibrium nature, and generates fractal geometries in space and time.

The term fractal was introduced by Mandelbrot and Van Ness [13, 14] to describe objects that display self-similarity. Since then data exploration in terms of fractal properties has found regular use in research areas as diverse as economics [15], engineering [16, 17], financial markets [18], geology [19, 20], hydrology [21, 22], medicine [23–25], physics [25], space science [26–28] and many more. Fractal properties in nature and human dynamics arguably serve to yield increased understanding and advancement in human society and stewardship.

Dynamic features in natural and social systems often show the scale independent behavior characteristic of fractals [19, 29, 30]. One can define a fractal set according to the following power law,

$$N \propto \frac{1}{r^D} \tag{1}$$

where N is the value of an element of the set with characteristic linear dimension r, and D is the associated fractal dimension. The fractal dimension is a real number. It is equivalent to an Euclidean dimension when it has integer values. Not all fractals are self-similar according to the definition of Eq. (1); when the scale invariance depends on the direction, one speaks of self-affinity. A self-affine fractal may be defined in terms of a two-dimensional space (x,y) where the function $f(rx,r^H y)$ is statistically similar to $f(x,y)$. That is, self-affine fractals demonstrate self-similarity only if different coordinates are scaled by different magnification factors. The quantity r is a scaling factor and H is the dimensionless Hurst exponent, also known as the Hausdorff dimension, which is related to the fractal dimension (D) by $H = 2 - D$ [31]. In the following sections we will provide an overview of various approaches to estimate fractal properties, including the fractal dimension, of various physical and social systems. These approaches use the tools of chaos, nonlinear dynamics, and statistical mechanics to demonstrate the wide applicability to extreme events and risk management.

2. Fractal methods

In the following subsections we will illustrate the use of various fractal methods in a range of nonequilibrium systems. Our main goal is to summarize how fractal methods have wide, interdisciplinary applications in systems that are open, dissipative, and whose source parameters and interactions are not well prescribed, and which have an unknown, but large, number of degrees of freedom.

2.1 Rescaled range (R/S) analysis

Range scaling analysis, also known as the R/S method, was developed by H.E. Hurst [32] to study time series in civil engineering whose underlying processes are independent, though not necessarily Gaussian.

Let us define a time series $x(t)$ over the interval T with values $x(t_0)$, $x(t_1)$, $x(t_2)$, ..., $x(t_n)$ where $t_0 = 0$, $t_1 = \tau$, $t_2 = 2\tau$, ... , $t_N = N\tau$; $N \in \mathbb{N}$. The time interval between measurements is τ, and $t_n = T$. The mean signal is

$$\bar{x}(T) = \frac{1}{T}\int_0^T x(t)dt \tag{2}$$

and signal variance is

$$V(T) = \frac{1}{T}\int_0^T [x(t) - \bar{x}(T)]^2 dt \tag{3}$$

with standard deviation

$$\sigma(T) = [V(T)]^{1/2} \tag{4}$$

A rescaled range is calculated by first normalizing the original data by subtracting the sample mean:

$$Z_r = x(t_r) - \bar{x}; r\epsilon\mathbb{N} \tag{5}$$

Then an integrated, or cumulative, time series, y, is derived from Eq. (5),

$$y_l = \sum_{i=1}^{l} Z_i; l = 2, 3, \dots, N, \tag{6}$$

and an adjusted range, $R(T)$, is formed in terms of the maximum minus the minimum value of the cumulative series, Y,

$$R(T) = \sup(y_1, y_2, \dots, y_T) - \inf(y_1, y_2, \dots, y_T). \tag{7}$$

The rescaled range, R/S, is $R(T)/\sigma(T)$. It scales with respect to T, by a power law.

$$(R/S)_T \propto T^H \tag{8}$$

where H is the Hurst exponent [32]. **Figure 1** shows a magnetogram (top) from the experiment described by [33]. The magnetic field was measured by the author, at a 10-second cadence, near the Hartebeesthoek Satellite Tracking Station and Radio Astronomy Observatory (HBS), near Pretoria, South Africa on March 1, 1994, using a Geometrics G-856 magnetometer. The bottom subplot shows the log-log plot of the rescaled range (R/S), the slope of which yields a scaling exponent H = 0.39 ± 0.01.

2.2 Spectral analysis

Fourier transforms are an efficient technique for decomposing a signal from the temporal or spatial domain into the frequency domain. Fractal scaling can be

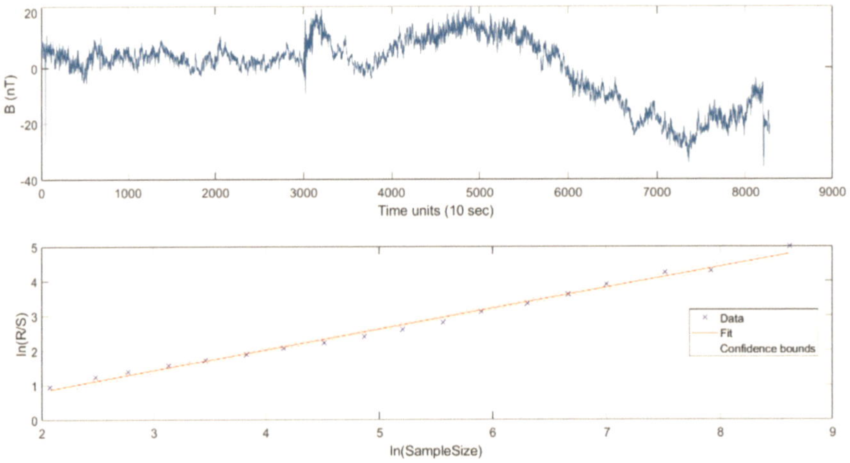

Figure 1.
(Top) Total magnetic field measured at Hartebeesthoek (HBS) in March 1994. (Bottom) Log-log plot of the corresponding rescaled range (R/S). The blue crosses show the data and the red line is the best-fit, 95% confidence intervals.

examined, under certain conditions, using a power spectral density (PSD) analysis. If the PSD, given as a function of frequency, follows a power-law, in other words $P(\nu)$ $\nu^{-\beta}$. The regularity of the scaling exponent is connected with the physical nature of the signal. That is, the slope β obtained from the plot of power spectral density against frequency provides a measure of the correlation level of the signal. When $-1 < \beta < 1$, the signal is termed a fractional Gaussian noise (fGn). fGn can be obtained as successive increments of a signal that displays fractional Brownian motion (fBm), which is a signal where both the real and imaginary components of the Fourier amplitudes are Gaussian-distributed random variables [34, 35]. The mean of the signal Fourier amplitudes $\overline{\varphi(\nu)} = 0$ and $\overline{\varphi(\nu)\varphi(\nu')^*} = P(\nu)\delta(\nu - \nu')$. This means that for $\beta = 2$, fBm is the same as a random walk and the power law spectrum varies as an inverse square. When the scaling exponent is above $\beta = 2$, the signal is called persistent, because if the data at some point have $x(t_{i+1}) > x(t_i)$, for instance, then the probability is greater than 0.5 that $x(t_{i+2}) > x(t_{i+1})$. Signals with exponents below 2 are called antipersistent because if $x(t_{i+1}) > x(t_i)$, the probability is greater than 0.5 that $x(t_{i+2}) < x(t_{i+1})$. The special case where $\beta = 2$ ($H = 0.5$) indicates Brownian motion. When the signal is fBm, it exhibits power-law scaling with slope $1 < \beta < 3$. In this case the signal is nonstationary but has stationary increments over a range of scales. For fBm one can generally relate the power-spectral scaling index to the Hurst index, introduced in the previous section:

$$\beta = 2H + 1. \tag{9}$$

The blue crosses in **Figure 2** show the PSD for the HBS data shown before. This yields a linear best-fit fractal scaling exponent for all the scales of $\beta = 2.21 \pm 0.10$ ($H = 0.61 \pm 0.05$). This best-fit is not shown in the figure because there is a crossover region in the figure differentiating two different scaling regimes; the best-fit straight line for all frequencies, corresponding to a single monofractal scaling exponent, does not capture the statistics evident in the PSD, particularly at the lower frequencies. Instead, we plot the two red curves which show the best-fits, with the 95% confidence

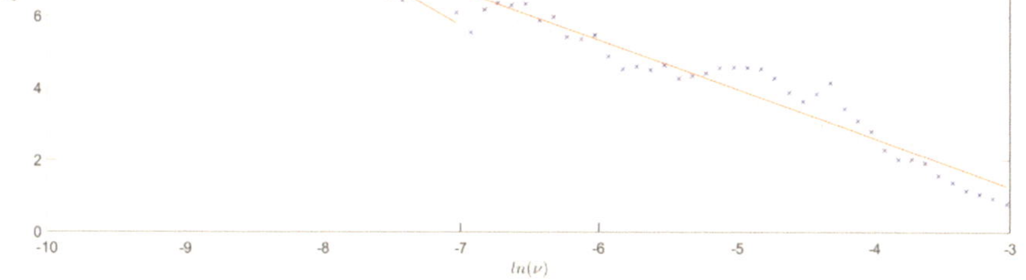

Figure 2.
The power spectral density versus frequency for the Hartebeesthoek (HBS) data (blue crosses). The red lines show the best-fit, 95% confidence intervals, for low frequency and higher frequency ranges. For low frequencies, $\beta = 2.20$ ($H = 0.60$) and for the higher frequencies the best-fit is $\beta = 1.43$ ($H = 0.22$).

intervals, for low frequency and higher frequency ranges on each side of the cross-over. For low frequencies, $\beta = 2.20 \pm 0.10$ ($H = 0.60 \pm 0.05$) and for the higher frequencies the best-fit yields $\beta = 1.43 \pm 0.02$ ($H = 0.22 \pm 0.01$). These results show what is often observed in dynamical data from different research areas, viz. that the correlations of recorded data often do not follow the same scaling law for all time scales [36, 37]. This can be because the statistical nature of the data is more complex than a simple fBm [38]. However, in the case illustrated in **Figure 2** it may be the result of nonstationarity of the data. The weakness inherent in spectral analysis is related to the problem of data nonstationarity [39]. Nonstationarity means that statistical properties are not constant through the signal, and traditional analysis methods that assume stationarity cannot be used, except with extreme care.

Although ubiquitous in science and engineering applications, Fourier analysis is often not suitable for the scrutiny of nonlinear and nonstationary data, which is precisely what is so common in the real world [40, 41]. Fourier techniques, without appropriate preconditioning of data, can yield spurious results for fractal properties. Thus, it behooves the researcher, when studying complex phenomena, whether that be turbulence or any other such systems with intermittent and extreme events, to carefully consider the nature of the data before choosing a method to study fractal properties, since data may feature complex statistics, including multifractional, multifractal and other models [27, 42–44].

2.3 Higuchi method

Higuchi [45] developed a widely applied time-domain technique to determine fractal properties of nonstationary physical data that is complex and non-periodic [46–48]. It is deemed a good tool to accurately estimate the fractal dimension and is linked to the fractal dimension derived from Fourier spectra when the signal is a fBm [49]. The Higuchi method was first applied to study large-scale turbulent fluctuations of magnetic fields in space. In that context it allowed modification of existing methods to study the fluctuation properties of turbulent plasma properties beyond the inertial range [50]. It is simple to implement, efficient, and can rapidly achieve accurate and stable values of fractal dimension. Importantly, this is true even in noisy, nonstationary data [51].

The Higuchi method takes a time series, with properties already defined in Eqs. (1)-(4) and, from this series, obtains X_k^m, defined as:

$$X_k^m : x(m), x(m+k), x(m+2k), \dots, x\left(m + \left[\frac{N-k}{k}\right] \cdot k\right), \tag{10}$$

Here [] represents the integer part of the enclosed value. The integer $m = 1, 2, \dots, k$ is the zero-point time and k is the time interval, with $k = 1, \dots, k_{max}$; k_{max} is a free tuning parameter. This means that given time interval equal to k, spawns k-sets of new time series with lengths defined by:

$$L_m(k) = \frac{\left\{ \left(\sum_{i=1}^{\left[\frac{N-m}{k}\right]} \left| x(m+ik) - x(m+(i-1)\cdot k \right| \right) \frac{N-1}{\left[\frac{N-m}{k}\right]\cdot k} \right\}}{k}. \tag{11}$$

The length of the curve for the time interval k is then defined as the average over the k sets of $L_m(k)$, i.e. $L(k) = \langle L_m(k) \rangle$. In cases when this equation scales according

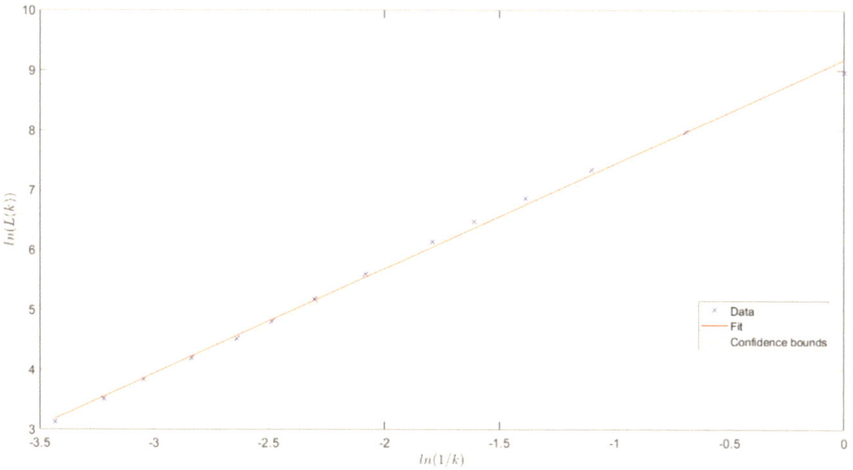

Figure 3.
Average curve length versus scale size, k, for the HBS time series yielding HFD = 1.75, and corresponding H = 0.25. The blue crosses show the data and the red line is the best-fit, with 95% confidence intervals.

to the rule $\boldsymbol{L(k)} \propto \boldsymbol{k}^{-\mathbf{HFD}}$, the time series behaves as a fractal with dimension *HFD*. A problem with the method, not apparently well-appreciated in the research community, is that a poor choice of the tuning parameter (k_{max}); can generate spurious results. Recent research has suggested various methodologies to choose an appropriate tuning parameter based on the input data [52–55]. We follow the methodology of [52], with k_{max} = 37 to derive the results in **Figure 3**, which yield $HFD = 1.75 \pm 0.02$, and since H = 2-HFD this gives $H = 0.25 \pm 0.02$.

2.4 Structure function analysis

Structure functions have traditionally been used in studies of plasma turbulence [56]. Compared to techniques such as Fourier analysis, structure functions can derive results with irregularly sampled data, and makes no assumptions about the missing data, such as stationarity. In terms of spatial fields, the structure function $S_p(x, r)$ of order p of a field $\tau(x)$ is given by:

$$S_p(\vec{x},\vec{r}) = \langle |\tau(\vec{x}+\vec{r}) - \tau(\vec{x})|^p \rangle \tag{12}$$

Here x is the position and r is a lag with respect to x, and the angled braces, $\langle \cdot \rangle$, indicate an ensemble average. This can easily be generalized to a time-series. Structure functions of high order (large values of p) yield information concerning the distribution of the large fluctuations of the spatial or temporal field. Smaller fluctuations generally dominate the lower-order structure functions.

If increments of the field $\tau(x + r) - \tau(x)$ are assumed to be statistically homogeneous, $S_p(x, r)$ is only a function of the lag r and may be written as $S_p(r)$. In this situation, the ergodic theorem implies that a spatial average can be used to estimate the ensemble average in the equation.

For white noise [57], the second-order structure function scales trivially in time as $S_2 \sim c$, a constant. In the case of diffusive processes, such as the well-known stochastic process of Brownian motion, the scaling obeys as $S_2 \sim \tau$. For fractional Brownian

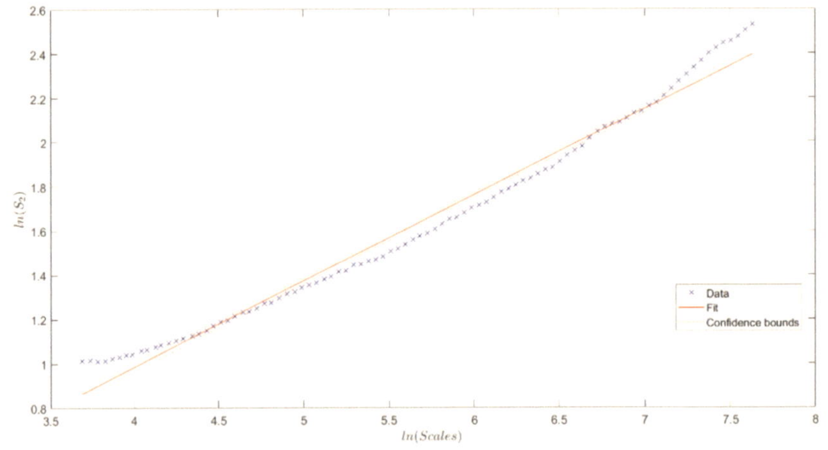

Figure 4.
Plot of second-order structure function (blue crosses) and best-fit red line with 95% confidence intervals for the Hartebeesthoek (HBS) data, yielding H = 0.20 ± 0.01.

motion (having white noise and Brownian motion as subclasses) the scaling obeys the fractal scaling law $S_2 \sim \tau^{2H}$, where H is the Hurst exponent. **Figure 4** shows the second-order structure function results, which yield H = 0.20 ± 0.01. The curve deviates significantly from a power-law at the smallest and largest scales. The Hurst exponent estimates can clearly be strongly influenced by the choice of the minimum and maximum fitting scale [58]. In addition, the length of the time-series can influence scaling estimates, thus finite sample properties of estimators should not be ignored [55, 58].

2.5 Detrended fluctuation analysis (DFA)

The final method we consider is detrended fluctuation analysis (DFA) which was initially introduced to study DNA nucleotides and variations in the human heart rate, which display the kind of long-range correlations typical of nonequilibrium dynamical systems [59, 60]. It has since then been employed in an enormous range of fields including economics, engineering, space science, and many others [61–64]. The method is a modified root mean squared analysis of a random walk designed specifically to be able to deal with nonstationarities in nonlinear data, and is considered to be among the most robust of statistical techniques designed to detect long-range correlations in time series [65–67]. Because DFA focuses on fluctuations around trend, rather than the signal range, it has been shown to be robust to the presence of trends [68] and nonstationary time series [37, 69].

Briefly, the methodology begins by removing the mean, $\bar{x}$, from the time series, x (t), and then integrating as in Eqs. (1)-(6). At this point the methodology deviates from the R/S analysis. Now the new time-series is divided into boxes of equal length, n. A trend, represented by a least-squares fit to the data, is removed from each box; the trend is typically a linear, quadratic, or cubic function. Box-size n has its abscissa denoted by $y_n(k)$. Next the trend is removed from the integrated time series, $y(k)$, by subtracting the local trend, $y_n(k)$, in each box.

For a given box size n, the characteristic size of the fluctuations, designated by the fluctuation function $F(n)$, is calculated as the RMS deviation between $y(k)$ and the trend in each box.

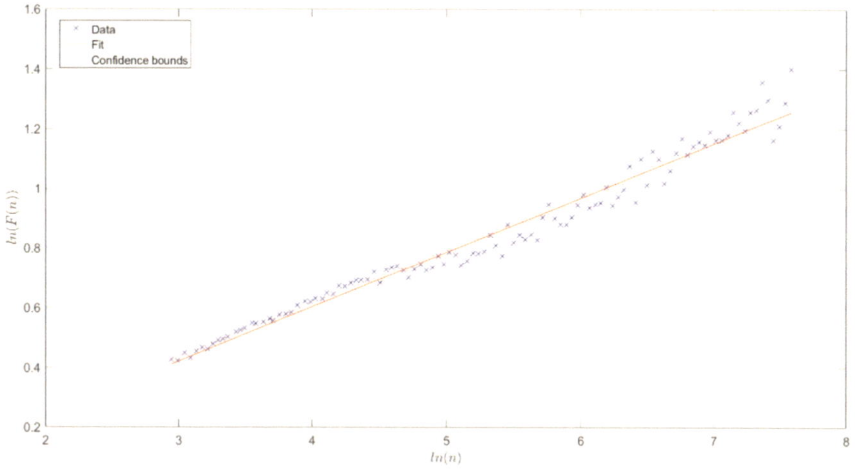

Figure 5.
Blue crosses show the fluctuation versus box size for the HBS data obtained by detrended fluctuation analysis. The best-fit red line with 95% confidence intervals has a slope corresponding to H = 0.19 ± 0.01.

$$F(n) = \sqrt{\frac{1}{N}\sum\nolimits_{k=1}^{N}\left[y(k) - y_n(k)\right]^2} \tag{13}$$

The calculation in Eq. (13) is finally performed over all time scales (box sizes), to potentially establish a power-law scaling between $F(n)$ and n, viz.

$$F(n) \propto n^{\alpha} \tag{14}$$

Here the α parameter is a scaling exponent. Given a fractional Brownian motion series, the scaling exponent α = H – the familiar Hurst exponent. **Figure 5** shows the fluctuation function results, which yield $H = 0.19 \pm 0.02$. At the largest scales there is considerable larger standard deviations for $F(n)$, than, for instance, compared to the R/S results in **Figure 1**. However, because of difference in methodology and finite sample effects the R/S estimate yields much larger Hurst exponent.

3. Summary and conclusion

In this paper we have looked at the use cases, in multiple fields of research, for various estimators of fractal long-range dependence. Analysis of data in terms of complexity science can help understand systems with natural hazards, and also human social and economic systems, where disruptive, extreme events result from emergent properties. This is because the value of system parameters at a certain time is related not just to the immediate past, but also to fluctuations in the remote past. Fractal analysis can thus allow insights into the origin and development of extreme events, and provide limits for theoretical models. The main goal of this paper has been to share and expand the scope of some of the fractal methods commonly used in complex systems studies. A proper appreciation of the strengths and limitations of fractal methods can improve the assessment and analysis of risks associated with systems that exhibit extreme events.

Method	Hurst exponent, H
R/S	0.39 ± 0.01
Spectral	$0.60 \pm 0.05/0.22 \pm 0.01$
Higuchi	0.25 ± 0.02
Structure Function	0.20 ± 0.01
DFA	0.19 ± 0.02

Table 1.
Comparison of Hurst exponent from different fractal methods.

The different classical estimators of fractal scaling behavior we considered were the R/S, spectral, Higuchi, structure function, and DFA methods. **Table 1** gives a comparison of the fractal scaling exponent results from the five different methods that we considered for the analysis of a short time series (8500 data points) of a ground magnetic field measured by the author near the Hartebeesthoek Satellite Tracking Station and Radio Astronomy Observatory (HBS), near Pretoria, South Africa on March 1, 1994.

The results we have obtained show how different Hurst scaling values may be obtained by employing different fractal methods on the same input data. The PSD analysis appears to show two different scaling regimes and a crossover (**Figure 2**), which can arise due to nonstationarity in the data. This illustrates the importance of determining first whether the data to be studied are stationary or nonstationary. Techniques, such as R/S and spectral analysis assume stationary data and can produce spurious and misleading results if applied without care on nonstationary data. This is evident in the large difference between the estimate for the Hurst exponent for the R/S method, compared to the other methods.

Another issue, demonstrated in **Figure 4**, relates to the length of the data series under investigation and other finite sample properties of estimators. **Figure 4** shows the second-order structure function results can deviate significantly from a power-law at the smallest and largest scales. Thus, Hurst exponent estimates can be influenced by the choice of the minimum and maximum fitting scale.

Next, although a single scaling law is possible, over a wide or infinite range of scales, due to the system dynamics there can be a multitude of actual scaling exponents and scaling regimes. In this case the system is not simply fractal, but multifractal [70, 71]. In such a complex system phase transitions in the regulation behavior can be connected to changes in internal fractal dynamics. This supports various hypotheses that connect extreme events to dynamical phase transitions in out of equilibrium systems [9, 61, 72].

Although we have not undertaken to determine which methods are most accurate, the cautions we have highlighted are important for researchers in the field to consider. Various studies have done relevant comparative studies of the different methods. For instance, some authors have used Monte Carlo methods of very specific synthetically generated data to test method accuracies at capturing scaling exponents [55, 58]. Others have considered the effects of trends, the length of the series, periodicities, and other considerations that influence determination of scaling behaviors [64, 66–68, 73–75]. Finally, there are various other methods of fractal dimension calculation that we have not discussed in this paper. These include the basic box-counting, dispersional, Hall-Wood, semi-periodogram, discrete cosine transform, and wavelet estimators [44, 76–78].

Acknowledgements

Research reported in this publication was supported by the National Science Foundation under Award Number AGS-2053689, and the National Institute of General Medical Sciences of the National Institutes of Health under Award Number P20GM103499. The content is solely the responsibility of the authors and does not necessarily represent the official views of any funding agency.

Author details

James Wanliss[1,2]

1 College of Engineering, Anderson University, Anderson, SC, USA

2 Presbyterian College, Clinton, SC, USA

*Address all correspondence to: jwanliss@andersonuniversity.edu

References

[1] Watkins NW, Pruessner G, Chapman SC, et al. 25 years of self-organized criticality: Concepts and controversies. Space Science Reviews. 2016;**198**:3-44. DOI: 10.1007/s11214-015-0155-x

[2] Dubrulle B. Multi-fractality, universality and singularity in turbulence. Fractal and Fractional 6. 2022;**10**:613. DOI: 10.3390/fractal fract6100613

[3] Cavalcante HL, Oria M, Sornette D, Ott E, Gauthier DJ. Predictability and suppression of extreme events in a chaotic system. Physical Review Letters. 2013;**111**:198701

[4] Arendarczyk M, Kozubowski TJ, Panorska AK. Preparing students for the future: Extreme events and power tails. Journal of Statistics and Data Science Education. 2022:1-5. DOI: 10.1080/26939169.2022.2146613

[5] Moffatt HK, Shuckburgh E, editors. Environmental Hazards: The Fluid Dynamics and Geophysics of Extreme Events. World Scientific. Lecture Notes Series, Institute for Mathematical Sciences, National University of Singapore. 2011. DOI: 10.1142/7796

[6] Werner TR, Gubiec T, Kutner R, et al. Modeling of super-extreme events: An application to the hierarchical Weierstrass-Mandelbrot continuous-time random walk. The European Physical Journal Special Topics. 2012; **205**:27-52. DOI: 10.1140/epjst/e2012-01560-0

[7] Sharma AS, Sitnov MI, Papadopoulos K. Substorms as nonequilibrium transitions in the magnetosphere. Journal of Atmospheric and Solar - Terrestrial Physics. 2001;**63**:1399

[8] Sitnov MI, Sharma AS, Papadopoulos K, Vassiliadis D, Valdivia JA, Klimas AJ, et al. Modeling substorm dynamics of the magnetosphere: From self-organization and self-organized criticality to nonequilibrium phase transitions. Physical Review E. 2001;**65**:016116

[9] Wanliss JA, Dobias P. Space storm as a dynamic phase transition. Journal of Atmospheric and Solar-Terrestrial Physics. 2007;**69**:675-684. DOI: 10.1016/j.jastp.2007.01.001

[10] Alberti T, Consolini G, De Michelis P, Laurenza M, Marcucci MF. On fast and slow Earth's magnetospheric dynamics during geomagnetic storms: A stochastic Langevin approach. Journal of Space Weather and Space Climate. 2018; **8**:A56. DOI: 10.1051/swsc/2018039

[11] Tsallis C. Possible generalization of Boltzmann-Gibbs statistics. Journal of Statistical Physics. 1988;**52**:479-487. DOI: 10.1007/BF01016429

[12] Tsallis C. On the foundations of statistical mechanics. The European Physical Journal Special Topics. 2017; **226**:1433-1443

[13] Mandelbrot B. How long is the coast of Britain? Statistical self-similarity and fractional dimension. Science. 1967;**15**: 636-638

[14] Mandelbrot B, Van Ness JW. Fractional Brownian motions: Fractional noises and applications. SIAM Reveiw 10. 1968;**4**:422-437

[15] Fama Eugene F, Cochrane John H, Moskowitz Tobias J. In: Fama F, editor. The Fama Portfolio: Selected Papers of Eugene. Chicago: University of Chicago

Press; 2021. DOI: 10.7208/9780226426983

[16] Li G, Zhang K, Gong J, Jin X. Calculation method for fractal characteristics of machining topography surface based on wavelet transform. Procedia CIRP. 2019;**79**:500-504. DOI: 10.1016/j.procir.2019.02.109

[17] Yang X, Xiang Y, Jiang B. On multi-fault detection of rolling bearing through probabilistic principal component analysis denoising and Higuchi fractal dimension transformation. Journal of Vibration and Control. May 2022;**28** (9-10):1214-1226. DOI: 10.1177/1077546321989527

[18] Wątorek M, Drożdż S, Kwapień J, Minati L, Oświęcimka P, Stanuszek M. Multiscale characteristics of the emerging global cryptocurrency market. Physics Reports. 2021;**901**(2021):1-82. DOI: 10.1016/j.physrep.2020.10.005

[19] Turcotte DL. Fractals, chaos, self-organized criticality and tectonics. Terra Nova. 1992;**4**(1):4-12. DOI: 10.1111/j.1365-3121.1992.tb00444.x

[20] Ranguelov B, Ivanov Y. Fractal properties of the elements of plate tectonics. Journal of Mining and Geological Sciences. 2017;**60**(1):83-89

[21] Zuo R, Cheng Q, Agterberg FP, Xia Q. Application of singularity mapping technique to identify local anomalies using stream sediment geochemical data, a case study from Gangdese, Tibet, western China. Journal of Geochemical Exploration. 2009; **101**(3):225-235. DOI: 10.1016/j.gexplo.2008.08.003

[22] Koutsoyiannis D. Time's arrow in stochastic characterization and simulation of atmospheric and hydrological processes. Hydrological Sciences Journal. 2019;**64**:1013-1037

[23] Mitsutake G, Otsuka K, Oinuma S, Ferguson I, Cornélissen G, Wanliss J, et al. Does exposure to an artificial ULF magnetic field affect blood pressure, heart rate variability and mood? Biomedicine & Pharmacotherapy. 2004; **58**:S20-S27. DOI: 10.1016/S0753-3322(04)80004-0

[24] Grizzi F, Castello A, Qehajaj D, Russo C, Lopci E. The complexity and fractal geometry of nuclear medicine images. Molecular Imaging and Biology. 2019;**21**(3):401-409

[25] Wang W, Moore MA, Katzgraber HG. Fractal dimension of interfaces in Edwards-Anderson spin glasses for up to six space dimensions. Physical Review E. 2018;**97**(3):032104

[26] Wanliss JA, Reynolds MA. Measurement of the stochasticity of low-latitude geomagnetic temporal variations. Annales de Geophysique. 2003;**21**:2025. DOI: 10.5194/angeo-21-2025-2003

[27] Cersosimo DO, Wanliss JA. Initial studies of high latitude magnetic field data during different magnetospheric conditions. Earth, Planets and Space. 2007;**59**:39-43. DOI: 10.1186/BF03352020

[28] Martin P, Turner NE, Wanliss J. Timing of substorm related auroral oscillations. Earth, Planets and Space. 2009;**61**(5):615-619. DOI: 10.1186/BF03352935

[29] Dobias P, Wanliss JA. Intermittency of storms and substorms: Is it related to the critical behaviour? Annales Geophysicae. 2009;**27**:2011-2018. DOI: 10.5194/angeo-27-2011-2009

[30] Dobias P, Wanliss JA. Fractal properties of conflict in Afghanistan revisited. Journal of Battlefield Technology. 2012;**15**(3):31-36

[31] da Silva G, Turcotte DL. A comparison between Hurst and Hausdorff measures derived from fractional time series. Chaos, Solitons & Fractals. 1994;**4**(12):2181-2192

[32] Hurst HE. The long-term storage capacity of reservoirs. Transactions of the American Society of Civil Engineers. 1951;**116**:770-799. DOI: 10.1061/TACEAT.0006518

[33] Wanliss JA, Antoine LAG. Geomagnetic micropulsations: Implications for high resolution aeromagnetic surveys. Exploration Geophysics. 1995;**26**(4):535-538. DOI: 10.1071/EG995535

[34] Hergarten S. Self-organized Criticality in Earth Systems. Vol. 49. Berlin, Heidelberg, New York: Springer Academic Press, Springer-Verlag; 2002

[35] Li M. Fractal time series - A tutorial review. Mathematical Problems in Engineering. 2010;**2010**:26. Article ID 157264. DOI: 10.1155/2010/157264

[36] Kantelhardt JW, Koscielny-Bunde E, Rego HH, Havlin S, Bunde A. Detecting long-range correlations with detrended fluctuation analysis. Physica A. 2001;**295** (3–4):441-454

[37] Kantelhardt JW, Zschiegner SA, Koscielny-Bunde E, Havlin S, Bunde A, Stanley HE. Multifractal detrended fluctuation analysis of nonstationary time series. Physica A. 2002;**316**:87

[38] Setty VA, Sharma AS. Characterizing detrended fluctuation analysis of multifractional Brownian motion. Physica A: Statistical Mechanics and its Applications. 2015;**419**:698-706. DOI: 10.1016/j.physa.2014.10.016

[39] Brillinger DR. Fourier analysis of stationary processes. Proceedings of the IEEE. 1974;**62**:1628-1643

[40] Wanliss JA, Dobias P. Dealing with non-stationarities in violence data using empirical mode decomposition. Journal of Battlefield Technology. 2013;**16**(2): 9-12. DOI: 10.3316/informit.4129650 98010362

[41] Singh P, Joshi SD, Patney RK, Saha K. The Fourier decomposition method for nonlinear and non-stationary time series analysis. Proc. R. Soc. A.: Mathematical, Physical and Engineering Sciences. 2017. DOI: 10.1098/rspa.2016.0871

[42] Yu Z-G, Anh V, Wang Y, Mao D, Wanliss J. Modeling and simulation of the horizontal component of the geomagnetic field by fractional stochastic differential equations in conjunction with empirical mode decomposition. Journal of Geophysical Research. 2010;**115**:A10219. DOI: 10.1029/2009JA015206

[43] Lévy Véhel J. Beyond multifractional Brownian motion: New stochastic models for geophysical modelling. Nonlinear Processes in Geophysics. 2013;**20**: 643-655. DOI: 10.5194/npg-20-643-2013

[44] Maruyama F, Kai K, Morimoto H. Wavelet-based multifractal analysis on a time series of solar activity and PDO climate index. Advances in Space Research. 2017;**60**(6):1363-1372

[45] Higuchi T. Approach to an irregular time series on the basis of the fractal theory. Physica D. 1988;**31**:277-283

[46] Esteller R, Vachtsevanos G, Echauz J, Litt B. A comparison of

waveform fractal dimension algorithms. IEEE Transactions on Circuits and Systems I: Fundamental Theory and Applications. 2001;**48**:177-183

[47] Salazar-Varas R, Vazquez RA. Time-invariant EEG classification based on the fractal dimension. In: Castro F, Miranda-Jiménez S, González-Mendoza M, editors. Advances in Computational Intelligence. MICAI. 2017. Lecture Notes in Computer Science. Vol. 10633. Cham: Springer; 2018. DOI: 10.1007/978-3-030-02840-4_26

[48] Yilmaz A, Unal G. Multiscale Higuchi's fractal dimension method. Nonlinear Dynamics. 2020;**101**: 1441-1455. DOI: 10.1007/s11071-020-05826-w

[49] Kalauzi A, Bojić T, Vuckovic A. Modeling the relationship between Higuchi's fractal dimension and Fourier spectra of physiological signals. Medical & Biological Engineering & Computing. 2012 Jul;**50**(7):689-699. DOI: 10.1007/s11517-012-0913-9

[50] Burlaga LF, Klein LW. Fractal structure of the interplanetary magnetic field. Journal of Geophysical Research. 1986;**91**(A1):347-350

[51] Liehr L, Massopust P. On the mathematical validity of the Higuchi method. Physica D: Nonlinear Phenomena. 2020;**402**:132265. DOI: 10.1016/j.physd.2019.132265

[52] Paramanathan P, Uthayakumar R. Application of fractal theory in analysis of human electroencephalographic Signals. Computers in Biology and Medicine. 2008;**38**(3):372-378

[53] Gomolka RS et al. Higuchi fractal dimension of heart rate variability during percutaneous auricular Vagus nerve stimulation in healthy and diabetic subjects. Frontiers in Physiology. 2018;**9**: 1162. DOI: 10.3389/fphys.2018.01162

[54] Wanliss J, Arriaza RH, Wanliss G, Gordon S. Optimization of the Higuchi method. International Journal of Research. -Granthaalayah. 2021;**9**(11): 202-213. DOI: 10.29121/granthaalayah.v9.i11.2021.4393

[55] Wanliss JA, Wanliss GE. Efficient calculation of fractal properties via the Higuchi method. Nonlinear Dynamics. 2022;**109**:2893-2904. DOI: 10.1007/s11071-022-07353-2

[56] Yu CX, Gilmore M, Peebles WA, Rhodes TL. Structure function analysis of long-range correlations in plasma turbulence. Physics of Plasmas. 2003;**10**: 2772. DOI: 10.1063/1.1583711

[57] Hida T, Kuo H, Potthoff J, Strei L. White Noise: An Infinite Dimensional Calculus. Netherlands: Springer; 2009

[58] Weron R. Estimating long-range dependence: Finite sample properties and confidence intervals. Physica A: Statistical Mechanics and its Applications. 2002;**312**(1-2):285-299. DOI: 10.1016/S0378-4371(02)00961-5

[59] Peng C, Buldyrev S, Havlin S, Simons M, Stanley H, Goldberger A. Mosaic organization of DNA nucleotides. Physical Review E. 1994;**49**(2)

[60] Peng C-K, Havlin S, Stanley HE, Goldberger AL. Quantification of scaling exponents and crossover phenomena in nonstationary heartbeat timeseries. Chaos. 1995;**5**(1):82-87

[61] Wanliss JA. Statistical precursors to space storm onset. In: Lui ATY, Kamide Y, Consolini G, editors. Multiscale Coupling of Sun-Earth

Processes. Elsevier; 2005. pp. 99-112. DOI: 10.1016/B978-044451881-1/50008-3

[62] Wanliss J. Efficiency of Price movements in futures markets. The Indian Economic Journal. 2020;**68**(2):193-206. DOI: 10.1177/0019466220966599

[63] Wanliss J, Dobias P. Detrended fluctuation analysis of combat in Afghanistan. Journal of Battlefield Technology. 2010;**13**(3):25-29. DOI: 10.3316/informit.482110479948687

[64] Wanliss JA, Shiokawa K, Yumoto K. Latitudinal variation of stochastic properties of the geomagnetic field. Nonlinear Processes in Geophysics. 2014; **21**:347-356. DOI: 10.5194/npg-21-347-2014

[65] Taqqu MS, Teverovsky V, Willinger W. Estimators for long-range dependence: An empirical study. Fractals. 1996;**3**(4):785-798

[66] Cannon MJ, Percival DB, Caccia DC, Raymond GM, Bassingthwaighte JB. Evaluating scaled windowed variance methods for estimating the Hurst coefficient of time series. Physica A. 1997;**241**(3):606-626

[67] Blok HJ. On the Nature of the Stock Market: Simulations and Experiments, PhD thesis. Canada: University of British Columbia; 2000

[68] Hu KP, Ivanov C, Chen Z, Carpena P, Stanley HE. Effect of trends on detrended fluctuation analysis. Physical Review E. 2001;**64**(1):011114

[69] Chen Z, Ivanov PC, Hu K, Stanley HE. Effect of nonstationarities on detrended fluctuation analysis. Physical Review E. 2002;**65**:041107

[70] Kantelhardt JW. Fractal and multifractal time series. In: Meyers RA, editor. Mathematics of Complexity and Dynamical Systems. New York: Springer; 2011

[71] Wanliss JA, Anh VV, Yu Z-G, Watson S. Multifractal modeling of magnetic storms via symbolic dynamics analysis. Journal of Geophysical Research. 2005. DOI: 10.1029/2004JA010996

[72] Wanliss JA, Muñoz V, Pastén D, Toledo B, Valdivia JA. Critical behavior in earthquake energy dissipation. The European Physical Journal B. 2017;**90**(9):167. DOI: 10.1140/epjb/e2017-70657-y

[73] Talkner P, Weber RO. Power spectrum and detrended fluctuation analysis: Application to daily temperatures. Physical Review E. 2000; **62**(1):150

[74] Delignieres D, Ramdani S, Lemoine L, Torre K, Fortes M, Ninot G. Fractal analyses for 'short' time series: A re-assessment of classical methods. Journal of Mathematical Psychology. 2006;**50**(6):525-544. DOI: 10.1016/j.jmp.2006.07.004

[75] Xu L, Ivanov PC, Hu K, Chen Z, Carbone A, Stanley HE. Quantifying signals with power-law correlations: A comparative study of detrended fluctuation analysis and detrended moving average techniques. Physical Review E. 2005;**71**(5):051101

[76] Schleicher D. Hausdorff dimension, its properties, and its surprises. The American Mathematical Monthly. 2007; **114**(6):509-528. Available from: http://www.jstor.org/stable/27642249

[77] Manimaran P, Panigrahi PK, Parikh JC. Multiresolution analysis of fluctuations in non-stationary time series through discrete wavelets. Physica A: Statistical Mechanics and its

Applications. 2009;**388**(12). DOI: 10.1016/j.physa.2009.02.011

[78] Gneiting T, Ševčíková H, Percival DB. Estimators of fractal dimension: Assessing the roughness of time series and spatial data. Statistical Science. 2012;**27**(2):247-277. Available from: http://www.jstor.org/stable/41714797

Chapter 3

Derivation and Integration on a Fractal Subset of the Real Line

Donatella Bongiorno

Abstract

Ordinary calculus is usually inapplicable to fractal sets. In this chapter, we introduce and describe the various approaches made so far to define the theory of derivation and integration on fractal sets. In particular, we study some Riemann-type integrals (the s-Riemann integral, the sHK integral, the s-first-return integral) defined on a closed fractal subset of the real line with finite and positive s-dimensional Hausdorff measure (s-set) with particular attention to the Fundamental Theorem of Calculus. Moreover, we pay attention to the relation between the s-Riemann integral, the sHK integral, and the Lebesgue integral with respect to the Hausdorff measure $\mathcal{H}^s$, respectively, and we give a characterization of the primitives of the sHK integral.

Keywords: Hausdorff measure, s-set, s-Riemann integral, sHK integral, s-derivatives

1. Introduction

For many years, it was thought that the structure of fractal sets had so many irregularities to render too difficult the definition of standard-type methods and techniques of ordinary calculus in such type of sets. In fact, for example, the usual derivative of the classical Lebesgue-Cantor staircase function is zero almost everywhere, and the usual Riemann integral of a function defined on a fractal set is undefined (see Refs. [1–3]). Therefore analysis of fractals has been studied by different methods such as harmonic analysis, stochastic processes, and fractional processes. Very recently, a non-Newtonian calculus on fractal sets of the real line that starts by elementary non-Diophantine arithmetic operations of a Burgin type was formulated by M. Czachor (see Ref. [4]).

In this chapter, we present a method of standard calculus for fractal subsets of the real line, that was independently formulated by various authors (see Refs. [5–9]). Such formulation is aimed at those self-similar fractal subsets of the real line with finite and positive s-dimensional Hausdorff measure, briefly called s-sets. Moreover, it differs from the classical one, for the use of the Hausdorff measure instead of the natural distance. The idea of replacing the usual distance with the Hausdorff measure was used for the first time in the definition of s-derivative given in 1991 by De Guzmann, Martin, and Reyes (see Ref. [5]), in order to study the problem of existence and uniqueness of the solutions of ordinary differential equations in which the independent variable takes value in a fractal set. Later, in 1998, this concept was taken up by Jung and Su in Ref. [6] to define an integral of the Riemann type called s-integral.

Moreover, Parvate and Gangal in Ref. [7], independently by Jung and Su, introduced an integral of Riemann type on an s-set of the real line, called F^s-integral. Such integration processes were defined as the classical Riemann integral but with the Hausdorff measure and the mass function, respectively, taking over the role of the distance. Since the Hausdorff measure and the mass function are proportional (see Ref. [7], Section 4), it follows that Jiang and Su in Ref. [6] and Parvate and Gangal in Ref. [7] defined, independently, the same integral. In this chapter, we call it the s-Riemann integral.

Both authors proved a version of the Fundamental Theorem of Calculus. About this, we recall that, in the real line, such fundamental theorem states that: *if* $f : [a, b] \to \mathbb{R}$ *is differentiable on* $[a, b]$, *then the function* $f'(x)$ *is integrable (in some sense) on* $[a, b]$ *and* $\int_a^b f'(t)dt = f(b) - f(a)$.

Unfortunately, as it happens in the real line, the s-Riemann integral is not the best integral for the formulation of the Fundamental Theorem of Calculus. This was the reason that motivated Bongiorno and Corrao in Ref. [8] to define an Henstock-Kurzweil integration process on an s-set of the real line and to formulate in Ref. [9] the best version of the Fundamental Theorem of Calculus on such s-sets. We recall that, in the real line, it is the Henstock-Kurzweil integral that, solving the problem of primitives, provides the best version of the Fundamental Theorem of Calculus (see Ref. [10]). Later and independently, also Golmankhaneh and Baleanu extended, in Ref. [11], the s-Riemann integral by introducing an integral of the Henstock–Kurzweil type. However, no version of the Fundamental Theorem of Calculus is proved in Ref. [11].

More precisely, the integral introduced in Ref. [8] is based on the use of the Hausdorff measure instead of the notion of the classical distance, as it was already done by Jung and Su in Ref. [6]. On the other hand, the mass function, instead of the classical distance, is used in Ref. [11], as it was already done by Parvate and Gangal in Ref. [7]. Precisely, Golmankhaneh and Baleanu have revised the notion of mass function given in Ref. [7] by defining a special mass function through the use of a gauge function previously introduced in Ref. [12]. However, since the Hausdorff measure and the mass function are proportional, without loss of generality, we infer that the two integrals coincide and we call the resulting integral the *sHK* integral. Moreover, following the literature, we prefer the use of the Hausdorff measure instead of the mass function to define the *sHK* integral. Finally, through the characterization of the *sHK* primitives, a descriptive definition of the *sHK* integral is given.

All sections of this chapter are related to each other; moreover, in order not to burden the reader, some proofs are reported in Appendix A, while Appendix B contains a brief history of the Henstock-Kurzweil integral. Particular attention is paid to the formulation of the best version of the Fundamental Theorem of Calculus. Afterward, the last section of this chapter is devoted to the formulation of a new integration process. Precisely an integral of the first return-type, called s-first-return integral, is defined. The idea of the first-return technique comes from the Poincaré first-return map of differentiable dynamics, and it was already used in differentiation and in integration theory (see Refs. [13–15]).

2. The s-derivatives

Let $\mathcal{L}(\cdot)$ be the usual Lebesgue measure, and let E be a compact subset of the real line. Given s, with $0 < s \leq 1$, let

$$\mathcal{H}^s(E) = \lim_{\delta\to 0} \inf \left\{ \sum_{i=1}^{\infty} \mathcal{L}(A_i)^s : E \subset \bigcup_{i=1}^{\infty} A_i,\ \mathcal{L}(A_i) \le \delta \right\} \tag{1}$$

be the *s-dimensional Hausdorff measure* of E.

Recall that $\mathcal{H}^s(\cdot)$ is a Borel regular measure and that the unique number s for which $\mathcal{H}^t(E) = 0$ if $t > s$ and $\mathcal{H}^t(E) = \infty$ if $t < s$ is called the Hausdorff dimension of E.

Definition 2.1. Let E be a compact subset of the real line and let $0 < s < 1$. E is called an s-set if it is measurable with respect to the s-dimensional Hausdorff measure $\mathcal{H}^s$ (briefly $\mathcal{H}^s$-measurable) and $0 < \mathcal{H}^s(E) < \infty$.

Therefore $\mathcal{H}^s$ is a Radon measure on each s-set (see Ref. [16]).

From now on we will denote by $E \subset [a, b]$ an s-set of the real line, by $a = minE$ and by $b = maxE$.

Definition 2.2. For $x, y \in E$ we set

$$d(x,y) = \begin{cases} \mathcal{H}^s([x,y] \cap E), & \text{if } x < y; \\ \mathcal{H}^s([y,x] \cap E), & \text{if } y < x. \end{cases} \tag{2}$$

Proposition 2.1. The function $(x, y) \rightsquigarrow d(x, y)$ from $E \times E \to \mathbb{R}^+ \cup \{0\}$ is a metric, and the space (E, d) is a complete metric space.

Proposition 2.2. The topology of the metric space (E, d) coincides with the topology induced on E by the usual topology of $\mathbb{R}$.

Definition 2.3. Let $F : E \to \mathbb{R}$ and let $x_0 \in E$. The *s-derivatives* of F on the left and on the right at the point x_0 are defined, respectively, as follows:

$$F'^{-}_{s}(x_0) = \lim_{\substack{x \to x_0^- \\ x \in E}} \frac{F(x_0) - F(x)}{d(x, x_0)} \tag{3}$$

$$F'^{+}_{s}(x_0) = \lim_{\substack{x \to x_0^+ \\ x \in E}} \frac{F(x) - F(x_0)}{d(x, x_0)} \tag{4}$$

when these limits exist.

We say that the *s-derivative* of F at x_0 exists if $F'^{-}_{s}(x_0) = F'^{+}_{s}(x_0)$ or if the s-derivative of F on the left (resp. right) at x_0 exists, and for some, $\varepsilon > 0$ we have $d(x_0, x_0 + \varepsilon) = 0$ (resp. $d(x_0 - \varepsilon, x_0) = 0$). The s-derivative of F at x_0, when it exists, will be denoted by $F'_s(x_0)$.

It is trivial to observe that by the previous definition, it follows the linearity of the s-derivative. Moreover.

Remark 2.1. If F is s-derivable at the point x_0, then F is continuous at x_0 according to the topology induced on E by the usual topology of $\mathbb{R}$.

Hereafter, for each interval $A \subset [a, b]$, we set $\widetilde{A} = A \cap E$.

Example 2.1. Let $E \subset [0, 1]$ be the ternary Cantor set. E is an s-set for $s = \log_3 2$ and $\mathcal{H}^s(E) = 1$ (see Ref. [2], Theorem 1.14). Moreover $\mathcal{H}^s \widetilde{\left[\frac{2}{3^n}, \frac{1}{3^{n-1}}\right]} = \frac{1}{2^n} = \mathcal{H}^s \widetilde{\left[0, \frac{1}{3^n}\right]}$, and $\mathcal{H}^s \widetilde{\left[\frac{2}{3^n}, \frac{7}{3^{n+1}}\right]} = \frac{1}{4^n} = \mathcal{H}^s \widetilde{\left[\frac{8}{3^{n+1}}, \frac{1}{3^{n-1}}\right]}$.

The function

$$F(x)=\begin{cases}\dfrac{(-2)^n}{n}\mathcal{H}^s\left[\widetilde{\dfrac{2}{3^n},x}\right], & x\in\left[\widetilde{\dfrac{2}{3^n},\dfrac{7}{3^{n+1}}}\right];\\ \dfrac{(-2)^n}{n}\mathcal{H}^s\left[\widetilde{\dfrac{8}{3^{n+1}},x}\right], & x\in\left[\widetilde{\dfrac{8}{3^{n+1}},\dfrac{1}{3^{n-1}}}\right];\\ 0, & x=0.\end{cases} \tag{5}$$

is s-derivable on E with

$$F_s'(x)=\begin{cases}\dfrac{(-2)^n}{n}, & x\in\left[\widetilde{\dfrac{2}{3^n},\dfrac{1}{3^{n-1}}}\right];\\ 0, & x=0.\end{cases} \tag{6}$$

Infact, for $x_0\in\left[\widetilde{\frac{2}{3^n},\frac{1}{3^{n-1}}}\right]$ it is:

$$F'^{-}_{s}(x_0)=\lim_{\substack{x\to x_0^-\\ x\in E}}\frac{F(x_0)-F(x)}{\mathcal{H}^s\left(\widetilde{[x,x_0]}\right)}=\frac{(-2)^n}{n}, \tag{7}$$

$$F'^{+}_{s}(x_0)=\lim_{\substack{x\to x_0^+\\ x\in E}}\frac{F(x)-F(x_0)}{\mathcal{H}^s\left(\widetilde{[x_0,x]}\right)}=\frac{(-2)^n}{n}. \tag{8}$$

So $F_s'(x_0)=(-2)^n/n$.

Moreover, for $x\in\left[\widetilde{\frac{2}{3^n},\frac{7}{3^{n+1}}}\right]$, it is

$$\left|\frac{F(x)-F(0)}{\mathcal{H}^s\left(\widetilde{[0,x]}\right)}\right|=\frac{2^n}{n}\frac{\mathcal{H}^s\left(\widetilde{\left[\frac{2}{3^n},x\right]}\right)}{\mathcal{H}^s\left(\widetilde{[0,x]}\right)}\leq\frac{2^n}{n}\frac{1}{4^n}2^n=\frac{1}{n} \tag{9}$$

and, for $x\in\left[\widetilde{\frac{8}{3^{n+1}},\frac{1}{3^{n-1}}}\right]$, it is

$$\left|\frac{F(x)-F(0)}{\mathcal{H}^s\left(\widetilde{[0,x]}\right)}\right|=\frac{2^n}{n}\frac{\mathcal{H}^s\left(\widetilde{\left[\frac{8}{3^{n+1}},x\right]}\right)}{\mathcal{H}^s\left(\widetilde{[0,x]}\right)}\leq\frac{2^n}{n}\frac{1}{4^n}2^n=\frac{1}{n} \tag{10}$$

Thus

$$F_s'(0)=\lim_{\substack{x\to 0\\ x\in E}}\frac{F(x)-F(0)}{\mathcal{H}^s\left(\widetilde{[0,x]}\right)}=0. \tag{11}$$

Definition 2.4. Let $F:\mathbb{R}\to\mathbb{R}$ be a function. A point $x\in\mathbb{R}$ is said to be a *point of change of* F if it is not a constant over any open interval (c,d) containing x. The set of all points of change of F is called the set of change of F, and it is denoted by $Sch(F)$.

Theorem 2.1 Let $F : [a, b] \to \mathbb{R}$ be a continuous function such that $Sch(F) \subset E$. Let us suppose that F is s-derivable at each point $x \in [a, b]$ and that $F(a) = F(b) = 0$. Then there exists a point $c \in E$ such that $F'_s(c) \geq 0$ and a point $d \in E$ such that $F'_s(d) \leq 0$.

In Ref. [7], it is possible to find the proof of the previous theorem, and an example that shows as the "fragmented nature" of the fractal set does not allow us to make an analog of Rolle's theorem. Furthermore, the law of mean and the Leibniz rule is also discussed in Ref. [7].

3. The s-Riemann integral

Definition 3.1. Let E be a closed s-set of the real line, $a = \min E$ and $b = \max E$. Let $A \subset [a, b]$ be an interval. We call *interval of* the set $\widetilde{A} = A \cap E$.

Definition 3.2. *A partition of E* is any collection $P = \left\{ \left(\widetilde{A}_i, x_i \right) \right\}_{i=1}^{p}$ of pairwise disjoint intervals $\widetilde{A}_i$ and points $x_i \in \widetilde{A}_i$ such that $E = \cup_i \widetilde{A}_i$.

Definition 3.3. Let $f : E \to \mathbb{R}$ be a function. It is said that f is s-Riemann integrable on E, if there exists a number I such that, for each $\varepsilon > 0$ there is a $\delta > 0$ with

$$\left| \sum_{i=1}^{p} f(x_i) \mathcal{H}^s \left(\widetilde{A}_i \right) - I \right| < \varepsilon \tag{12}$$

for each partition $P = \left\{ \left(\widetilde{A}_i, x_i \right) \right\}_{i=1}^{p}$ of E with $\mathcal{H}^s \left(\widetilde{A}_i \right) < \delta$, $i = 1, 2, \ldots, p$. The number I is called the s-Riemann integral of f on E and we set

$$I = (s) \int_E f(t) \, d\mathcal{H}^s(t). \tag{13}$$

The collection of all functions that are s-Riemann integrable on E will be denoted by $sR(E)$.

Theorem 3.1 If $f : E \to \mathbb{R}$ is continuous on E with respect to the induced topology, then $f \in sR(E)$.

Remark 3.1. The classical Riemann integral properties, such as linearity, additivity with respect to integrating domain and the mean-value theorem for integrals hold too for this new integral. See Refs. [6, 7] for more details.

Moreover, it is useful to remark that, as it happens in the real case, the Lebesgue integral of f with respect to the Hausdorff measure $\mathcal{H}^s$, here denoted by $(L) \int_E f(t) \, d\mathcal{H}^s(t)$, includes the s-Riemann integral.

Theorem 3.2 If $f \in sR(E)$ then f is Lebesgue integrable on E with respect to the Hausdorff measure $\mathcal{H}^s$, and

$$(L) \int_E f(t) \, d\mathcal{H}^s(t) = (s) \int_E f(t) d\mathcal{H}^s(t). \tag{14}$$

The proof can be found in Appendix A.

3.1 The Fundamental Theorem of Calculus

Definition 3.4. Let $F : E \to \mathbb{R}$ be a function. We say that F is $\mathcal{H}^s$-absolutely continuous on E if $\forall \varepsilon > 0 \exists \delta > 0$ such that

$$\sum_{k=1}^{n} |F(b_k) - F(a_k)| < \varepsilon \tag{15}$$

whenever $\sum_{k=1}^{n} \mathcal{H}^s\left[\widetilde{(a_k, b_k)}\right] < \delta$, with $a_k, b_k \in E,\ k = 1, \dots, n,$ and $a_1 < b_1 \le a_2 < b_2 \le \dots \le a_n < b_n$.

Jiang and Su in Ref. [6] announced, without proof, the following version of the Fundamental Theorem of Calculus:

Theorem 3.2 Let $f : E \to \mathbb{R}$ be a continuous function. If $F : E \to \mathbb{R}$ is $\mathcal{H}^s$-absolutely continuous on E with $F'_s(x) = f(x)$ $\mathcal{H}^s$- a.e. in E, then

$$(s)\int_E f(t)\, d\mathcal{H}^s(t) = F(b) - F(a). \tag{16}$$

On the other hand Parvate and Gangal in Ref. [7] proved the following version of the Fundamental Theorem of Calculus:

Theorem 3.3 If $F : \mathbb{R} \to \mathbb{R}$ is s-derivable on E, and if F'_s is continuous with $Sch(F) \subseteq E$, then

$$(s)\int_E F'_s(x)\, d\mathcal{H}^s(x) = F(b) - F(a). \tag{17}$$

Now, we give an example of a very simple function, that is not $\mathcal{H}^s$-absolutely continuous on a fractal set E and that satisfies the condition $Sch(F) \subsetneq E$ for which the Fundamental Theorem of Calculus fails.

Example 3.1. Let $E \subset [0, 1]$ be the classical Cantor set. Let F be a real function on $[0, 1]$ defined by

$$F(x) = \begin{cases} 0, & 0 \le x \le \frac{1}{3} \\ 3x - 1, & \frac{1}{3} < x < \frac{2}{3} \\ 1, & \frac{2}{3} \le x \le 1. \end{cases} \tag{18}$$

and let f be the restriction of F on the Cantor set E.

Since, the Hausdorff dimension of the Cantor set E is $s = \log_3 2$ and $\mathcal{H}^{\log_3 2}(E) = 1$ it is trivial to observe that:

- f is $\log_3 2$-derivable on E with $f'_{\log_3 2}(x) = 0, \forall x \in E$,
- $f'_{\log_3 2}$ is $\log_3 2$-Riemann integrable on E,

$$(\log_3 2)\int_E f'_{\log_3 2}(x)\, d\mathcal{H}^{\log_3 2}(x) = 0 \ne f(1) - f(0) = 1. \tag{19}$$

In the next section, we will prove a more general formulation of the Fundamental Theorem of Calculus on an s-set, that enclose Theorem 3.2 and Theorem 3.3.

4. The *sHK* integral

Definition 4.1. It is called *gauge* on E any positive real function δ defined on E.

Definition 4.2. It is called *partition* of E any collection $P = \left\{ \left(\widetilde{A}_i, x_i \right) \right\}_{i=1}^{p}$ of pairwise disjoint intervals $\widetilde{A}_i$ of E, and points $x_i \in \widetilde{A}_i, i = 1, \cdots, p,$ such that $E = \bigcup_i \widetilde{A}_i$.

Definition 4.3. Let $P = \left\{ \left(\widetilde{A}_i, x_i \right) \right\}_{i=1}^{p}$ be a partition of E. If δ is a gauge on E, then we say that P is a δ-fine *partition* of E whenever $\widetilde{A}_i \subseteq]x_i - \delta(x_i), x_i + \delta(x_i)[$, for $i = 1, 2, \ldots, p$.

Lemma 4.1. If δ is a gauge on E, then there exists a δ-fine partition of E.

This lemma, known in the literature as the Cousin lemma, is fundamental for the definition of an Henstock-Kurzweil type integral because it addresses the existence of δ-fine partitions. For completeness, we report the proof of this lemma in Appendix A, even if it is possible to find it in Ref. [8].

Now, let $P = \left\{ \left(\widetilde{A}_i, x_i \right) \right\}_{i=1}^{p}$ be a partition of E, let $f : E \to \mathbb{R}$ be a function and let

$$\sigma(f, P) = \sum_{i=1}^{p} f(x_i) \mathcal{H}^s \left(\widetilde{A}_i \right), \tag{20}$$

be the s-Riemann sum of f with respect to P.

Definition 4.4. We say that f is *sHK* integrable on E if there exists $I \in \mathbb{R}$ such that, for all $\varepsilon > 0$, there is a gauge δ on E with:

$$|\sigma(f, P) - I| < \varepsilon, \tag{21}$$

for each δ-fine partition $P = \left\{ \left(\widetilde{A}_i, x_i \right) \right\}_{i=1}^{p}$ of E.

The number I is called the *sHK* integral of f on E, and we write

$$I = (sHK) \int_E f \, d\mathcal{H}^s. \tag{22}$$

The collection of all functions that are *sHK* integrable on E will be denoted by $sHK(E)$.

Remark 4.1. Let us notice that the difference between the *sHK* integral (see Definition 4.4) and the s-Riemann integral (see Definition 3.3) is due to the fact that while in Definition 4.4 the gauge δ is a positive real function in Definition 3.3 it is a positive constant.

About this, remark that, if f is an *sHK* integrable function, but it is not s-Riemann integrable, then $\inf_{x \in E} \delta(x) = 0,$ for the gauge δ involved in the definition of the *sHK* integral. In fact, the condition $\inf_{x \in E} \delta(x) = \delta > 0$ would imply that the choice of points x_i inside the intervals $\widetilde{A}_i, i = 1, 2, \cdots, p,$ may be arbitrary; therefore f would be s-Riemann integrable.

Here we list some basic properties of the *sHK* integral:

a. the number I from Definition 4.4 is unique,

b. if $f, g \in sHK(E)$ therefore $f + g \in sHK(E)$ and

$$(sHK) \int_E (f + g)(t) \, d\mathcal{H}^s(t) = (sHK) \int_E f(t) \, d\mathcal{H}^s(t) + (sHK) \int_E g(t) \, d\mathcal{H}^s(t),$$

c. if $f \in sHK(E)$ and $k \in \mathbb{R}$, then $kf \in sHK(E)$ and

$$(sHK)\int_E kf(t)\,d\mathcal{H}^s(t) = k(sHK)\int_E f(t)\,d\mathcal{H}^s(t),$$

d. if $f, g \in sHK(E)$ with $f \leq g$, $\mathcal{H}^s$-almost everywhere on E, therefore

$$(sHK)\int_E f(t)\,d\mathcal{H}^s(t) \leq (sHK)\int_E g(t)\,d\mathcal{H}^s(t),$$

e. if $f \in sR(E)$, then $f \in sHK(E)$ and

$$(sHK)\int_E f(t)\,d\mathcal{H}^s(t) = (s)\int_E f(t)\,d\mathcal{H}^s(t),$$

f. if $f \in sHK(E)$ and $a = \min E < x < b = \max E$, then the function

$$F(x) = (sHK)\int_{\widetilde{[a,x]}} f\,d\mathcal{H}^s \tag{23}$$

is continuous and

$$(sHK)\int_E f\,d\mathcal{H}^s = (sHK)\int_{\widetilde{[a,x]}} f\,d\mathcal{H}^s + (sHK)\int_{\widetilde{[x,b]}} f\,d\mathcal{H}^s. \tag{24}$$

Just as in the case of the Henstock-Kurzweil integral (see Ref. [10]), there is a Cauchy criterion for a function to be *sHK* integrable on E. This is the content of the following theorem.

Theorem 4.1 A function $f: E \to \mathbb{R}$ is *sHK*-integrable on E if and only if for each $\varepsilon > 0$ there exists a gauge δ on E such that

$$|\sigma(f, P_1) - \sigma(f, P_2)| < \varepsilon, \tag{25}$$

for each pair P_1 and P_2 of δ-fine partitions of E.

The proof can be found in Ref. [8].

4.1 Relation with the Lebesgue integral

In this section, we prove that the *sHK* integral includes the Lebesgue integral with respect to $\mathcal{H}^s$. In order to do this, we recall the following Vitali-Carathéodory Theorem:

The Vitali-Carathéodory Theorem Let f be a real function defined on E. If f is Lebesgue integrable on E with respect to $\mathcal{H}^s$ and $\varepsilon > 0$, then there exist functions u and v on E such that $u \leq f \leq v$, u is upper semicontinuous and bounded above, v is lower semicontinuous and bounded below, and

$$(L)\int_E (v - u)\,d\mathcal{H}^s < \varepsilon. \tag{26}$$

Theorem 4.2 Let $f : E \to \mathbb{R}$ be a function. If f is Lebesgue integrable on E with respect to $\mathcal{H}^s$, then f is sHK-integrable on E and

$$(L)\int_E f\, d\mathcal{H}^s = (sHK)\int_E f\, d\mathcal{H}^s. \tag{27}$$

The proof can be found in Appendix A.

Now we give a simple example of an sHK integrable function which is not Lebesgue integrable with respect to the Hausdorff measure $\mathcal{H}^s$.

Example 4.1. Let $E \in [0, 1]$ be the classical Cantor set, and let $f : E \to \mathbb{R}$ be the function defined as follows

$$f(x) = \begin{cases} \dfrac{(-1)^{n+1}2^n}{n}, & \text{for } x \in \widetilde{\left[\dfrac{2}{3^n}, \dfrac{1}{3^{n-1}}\right]} \quad n = 1, 2, 3, \dots \\ 0, & \text{for } x = 0. \end{cases} \tag{28}$$

We show that $f \in sHK(E)$ where $s = \log_3 2$.

In order to do that, fixed $\varepsilon > 0$, we can find a gauge δ on E such that

- if $x \in E$ and $x \neq 0$, f is constant on $\left(\widetilde{x - \delta(x), x + \delta(x)}\right)$;
- $\delta(0) < \frac{1}{3^{n+1}}$ with $\frac{1}{n} < \varepsilon$.

Choose $k \in \mathbb{N}$ with $k > n + 1$ and set $c = \frac{1}{3^k}$.

Let us consider $P = \left\{\left(\widetilde{A_1}, x_1\right), \left(\widetilde{A_2}, x_2\right) \dots, \left(\widetilde{A_m}, x_m\right)\right\}$ a δ-fine partition of E such that $\widetilde{A_1} = \widetilde{[0, c]}$ and $\widetilde{A}_i \subseteq \widetilde{\left[\frac{2}{3^p}, \frac{1}{3^{p-1}}\right]}$ $(i = 2, \dots, m)$ for some $p \in \mathbb{N}$. Our choice of δ implies that $x_1 = 0$ and $\bigcup_{i=2}^m \widetilde{A}_i = \bigcup_{i=1}^k \widetilde{\left[\frac{2}{3^i}, \frac{1}{3^{i-1}}\right]}$.

$$\begin{aligned}(sHK)\int_{\widetilde{\left[\frac{2}{3^k},1\right]}} f\, d\mathcal{H}^s &= \sum_{i=1}^k (sHK)\int_{\widetilde{\left[\frac{2}{3^i},\frac{1}{3^{i-1}}\right]}} f\, d\mathcal{H}^s = \\ &= \sum_{i=1}^k \frac{(-1)^{i+1}\, 2^i}{i}\, \mathcal{H}^s\left(\widetilde{\left[\frac{2}{3^i}, \frac{1}{3^{i-1}}\right]}\right) = \\ &= \sum_{i=1}^k \frac{(-1)^{i+1}\, 2^i}{i} \cdot \frac{1}{2^i} = \sum_{i=1}^k \frac{(-1)^{i+1}}{i}.\end{aligned} \tag{29}$$

Therefore

$$(sHK)\int_{\widetilde{\left[\frac{2}{3^k},1\right]}} f\, d\mathcal{H}^s \sim \log 2. \tag{30}$$

Then

$$|\sigma(f,P) - \log 2| \le \left|f(0)\mathcal{H}^s(\widetilde{A_1})\right| + \left|\sum_{i=2}^{m} f(x_i)\mathcal{H}^s(\widetilde{A_i}) - (sHK)\int_{\widetilde{\left[\frac{2}{3^k},1\right]}} f\, d\mathcal{H}^s\right| + \left|(sHK)\int_{\widetilde{\left[\frac{2}{3^k},1\right]}} f\, d\mathcal{H}^s - \log 2\right| < \varepsilon. \tag{31}$$

In conclusion, we prove that f is not Lebesgue integrable on E with respect to $\mathcal{H}^s$, where $s = \log_3 2$. In fact, if f were Lebesgue integrable on E with respect to $\mathcal{H}^s$, $|f|$ would be Lebesgue integrable on E with respect to $\mathcal{H}^s$. But we have

$$(L)\int_E |f|\, d\mathcal{H}^s = \sum_{n=1}^{\infty} \frac{1}{n} = +\infty, \tag{32}$$

hence f is not Lebesgue integrable on E with respect to $\mathcal{H}^s$.

4.2 The Fundamental Theorem of Calculus

Definition 4.5. An interval (α, β) is said to be contiguous to a set E, if:

- $\alpha \in E$ and $\beta \in E$
- $(\alpha, \beta) \cap E = \emptyset$

In Ref. [9] we proved the following version of the Fundamental Theorem of Calculus.

Theorem 4.3 Let E be a closed s-set and let $\{(a_j, b_j)\}_{j\in\mathbb{N}}$ be the contiguous intervals of E. If $F : E \to \mathbb{R}$ is s-derivable on E and if $\sum_{j=1}^{\infty} |F(b_j) - F(a_j)| < +\infty$, then $F'_s \in sHK(E)$ and

$$(sHK)\int_E F'_s d\mathcal{H}^s = F(b) - F(a) - \sum_{j=1}^{\infty} (F(b_j) - F(a_j)). \tag{33}$$

Remark 4.2. By the previous theorem, it is possible to extend the versions of the Fundamental Theorem of Calculus given by Jung and Su (i.e., Theorem 3.2) and that given by Parvate and Gangal (i.e., Theorem 3.3).

Extension of Theorem 3.2.

Let E be a closed s-set and let $F : E \to \mathbb{R}$ be a function $\mathcal{H}^s$-absolutely continuous on E such that F'_s exists $\mathcal{H}^s$-almost everywhere in E. Then

$$(sHK)\int_E F'_s(t)\, d\mathcal{H}^s(t) = F(b) - F(a). \tag{34}$$

The proof of the Extension of Theorem 3.2 can be found in Ref. [9].

Extension of Theorem 3.3 If $F : \mathbb{R} \to \mathbb{R}$ is continuous and s-derivable on E and if $Sch(F) \subseteq E$, then

$$(sHK)\int_E F_s'(t)\, d\mathcal{H}^s(t) = F(b) - F(a). \tag{35}$$

Proof:

Condition $Sch(F) \subseteq E$ implies that F is constant on each contiguous interval (a_j, b_j) of E, then $F(a_j) = F(b_j)$ for $j \in \mathbb{N}$. Therefore $\sum_{j=1}^{\infty} |F(b_j) - F(a_j)| < +\infty$, and Theorem 4.3 can be applied.

Remark 4.3. If we assume, like in Theorem 3.3 that F_s' is continuous on E, then $F_s' \in sR(E)$ (Ref. [7], Theorem 39) and by (e) and by Extension of Theorem 3.3, we have that:

$$(s)\int_E F_s'(t)\, d\mathcal{H}^s(t) = F(b) - F(a). \tag{36}$$

Remark 4.4. The absolute convergence of the series $\sum_{j=1}^{\infty} (F(b_j) - F(a_j))$ is a necessary condition

- for the sHK integrability of F_s'
- for the validity of some formulation of the Fundamental Theorem of Calculus.

See Ref. [9] for more details.

4.3 The primitives

Bongiorno and Corrao in Ref. [17] introduced an Henstock-Kurzweil type integral defined on a complete metric measure space X endowed with a Radon measure μ and with a family $\mathcal{F}$ of μ-cells that enclose the sHK integral (see Ref. [17], Example 2.4). For such an integral, they proved an extension of the usual descriptive characterizations of the Henstock-Kurzweil integral on the real line in terms of ACG^* functions (see Ref. [17] and Appendix B). Here we report such descriptive characterization in the particular case in which $X = [a, b]$ is endowed with the Euclidean distance of $\mathbb{R}, E \subset [a, b]$ is an s-set and the family $\mathcal{F}$ of μ-cells coincides with the family $\mathcal{F}$ of all closed subintervals of $\widetilde{[a, b]}$.

Definition 4.6. A finite collection $\left\{\widetilde{A}_1, \dots, \widetilde{A}_m\right\}$ of pairwise disjoint elements of $\widetilde{[a, b]}$ is called a *division* of $\widetilde{[a, b]}$ if $\bigcup_{i=1}^{m} \widetilde{A}_i = \widetilde{[a, b]}$.

Definition 4.7. Let E be an s-set and let δ be a gauge on $\widetilde{[a, b]}$. A collection $P = \left\{\left(\widetilde{A}_i, x_i\right)\right\}_{i=1}^{m}$ of finite ordered pairs of intervals and points is said to be

- a *partial partition* of $\widetilde{[a, b]}$ if $\left\{\widetilde{A}_1, \dots, \widetilde{A}_m\right\}$ is a subsystem of a division of $\widetilde{[a, b]}$ and $x_i \in \widetilde{A}_i$ for $i = 1, 2, \dots, m$;
- *E-anchored* if the points $x_1, \dots, x_m$ belong to E.

Definition 4.8. Let $\pi : \mathcal{F} \to \mathbb{R}$ be a function. We say that π is an *additive function of interval* if for each $\widetilde{A} \in \mathcal{F}$ an for each division $\left\{\widetilde{A}_1, \dots, \widetilde{A}_m\right\}$ of $\widetilde{A}$ we have

$$\pi\left(\widetilde{A}\right) = \sum_{i=1}^{m} \pi\left(\widetilde{A}_i\right). \tag{37}$$

Definition 4.9. Let E be an s-set. Let π be a fixed additive function of interval. We say that π is AC^{Δ} on E if for $\varepsilon > 0$ there exists a gauge δ on E and a positive constant η such that the condition $\sum_{i=1}^{m} \mathcal{H}^s\left(\widetilde{A}_i\right) < \eta$, implies $\sum_{i=1}^{m} |\pi\left(\widetilde{A}_i\right)| < \varepsilon$, for each δ-fine E-anchored partial partition $P = \left\{\left(\widetilde{A}_i, x_i\right)\right\}_{i=1}^{m}$ of $\widetilde{[a,b]}$.

We say that π is ACG^{Δ} on $\widetilde{[a,b]}$ if there exists a countable sequence of s-sets $\{E_k\}_k$ such that $\bigcup_k E_k = \widetilde{[a,b]}$ and π is AC^{Δ} on E_k, for each $k \in \mathbb{N}$.

Theorem 4.4 Let $E \subset [a,b]$ be an s-set. A function $f : E \to \mathbb{R}$ is sHK integrable on E if and only if there exists an additive function of interval F that is ACG^{Δ} on $\widetilde{[a,b]}$ and s-derivable on $\widetilde{[a,b]}$ such that $F'_s(x) = f(x)$ $\mathcal{H}^s$-almost everywhere on $\widetilde{[a,b]}$.

Remark 4.5. An interested reader can find the proof of Theorem 4.4 in a more general version (i.e. for an Henstock-Kurzweil type integral defined on a complete metric measure space X endowed with a Radon measure μ) in Ref. [17].

5. The s-first-return integral

Darji and Evans in Ref. [15] have defined on the real line the first-return integral. The motivation that led the authors to define such a new integration process lies in the fact that the gauge function involved in the definition of the first-return integral is a constant like in the Riemann integral. Borrowing such an idea, in this section, we define a new integral on an s-set E, called the s-first-return integral. Such a new integral is different from the sHK integral because, in the definition of this new integral, the function $\delta : E \to \mathbb{R}^+$ is a positive constant, but the choice of points x_i is not arbitrary in $\widetilde{A}_i$, for $i = 1, 2, \cdots, p$.

Definition 5.1. We call *trajectory* on E any sequence Γ of distinct points of E dense in E. Given a trajectory Γ on E and given an interval, $\widetilde{A}$ we denote by $r\left(\Gamma, \widetilde{A}\right)$ the first element of Γ that belongs to $\widetilde{A}$.

Definition 5.2. Let $f : E \to \mathbb{R}$ and let Γ be a trajectory on E. We say that f is s-first-return integrable on E with respect to Γ if there exists a number $I \in \mathbb{R}$ such that, for all $\varepsilon > 0$, there is a constant $\delta > 0$ with:

$$\left| \sum_{i=1}^{p} f\left(r\left(\Gamma, \widetilde{A}_i\right)\right) \mathcal{H}^s\left(\widetilde{A}_i\right) - I \right| < \varepsilon \tag{38}$$

for each division $P = \left\{\widetilde{A}_i\right\}_{i=1}^{p}$ of E with $\mathcal{H}^s\left(\widetilde{A}_i\right) < \delta$.

The number I is called the s-first-return integral of f on E with respect to Γ, and we write

$$I = (sfr)_{\Gamma} \int_E f \, d\mathcal{H}^s. \tag{39}$$

The collection of all functions that are s-first-return integrable on E with respect to Γ will be denoted by $(sfr)_{\Gamma}(E)$.

Theorem 5.1 Let $f : E \to \mathbb{R}$ such that $f \in sR(E)$, therefore, there exists a trajectory Γ on E such that $f \in (sfr)_\Gamma(E)$ and

$$(sfr)_\Gamma \int_E f(t) d\mathcal{H}^s(t) = (s) \int_E f(t) d\mathcal{H}^s(t). \tag{40}$$

Theorem 5.2 There exists an s-derivative $f : E \to \mathbb{R}$ such that $f \notin (sfr)_\Gamma(E)$, for each trajectory Γ on E.

Remark 5.1. To prove Theorem 5.2 it is enough to consider the function $f(x) = F_s'(x)$ of the Example 2.1 and to show that $f \notin (sfr)_\Gamma(E)$, for a given trajectory Γ. This is equivalent to find, for each $M > 0$ and for each $\delta > 0$, a finite system of pairwise disjoint intervals $\widetilde{A}_i$, with $i = 1, 2, \dots, p$, such that $\mathcal{H}^s\left(\widetilde{A}_i\right) < \delta$, $\cup_i \widetilde{A}_i = E$ and

$$\sum_{i=1}^{p} f\left(r\left(\Gamma, \widetilde{A}_i\right)\right) \mathcal{H}^s\left(\widetilde{A}_i\right) > M. \tag{41}$$

See Ref. [18] for more details.

6. Conclusions

In this chapter we have developed a method of calculus on a closed fractal subset of the real line with finite and positive s-dimensional Hausdorff measure. Much of the development of such new calculus on s-sets is carried in analogy with the ordinary calculus with some differences due to the "fragmented nature" of fractal sets, like in the formulation of Rolle's theorem or in the formulation of the Fundamental Theorem of Calculus. Therefore, in order to give the best version of the Fundamental Theorem of Calculus (see Theorem 4.3), we have generalized the s-Riemann integral by defining the sHK integral and the s-first-return integral. Finally, by Theorem 5.2, we have noted that to obtain the best version of the Fundamental Theorem of Calculus, we need to consider an Henstock-Kurzweil type integral, i.e., an integral in which the gauge δ is not a constant.

Appendix A

Proof of Theorem 3.1.

By standard techniques, it follows that f is bounded and $\mathcal{H}^s$-measurable; then f is Lebesgue integrable with respect to the Hausdorff measure $\mathcal{H}^s$ (briefly, $\mathcal{H}^s$-Lebesgue integrable).

Given $\varepsilon > 0$, by Definition 3.3, there exists $\delta > 0$ such that

$$\left| \sum_{i=1}^{n} f(x_i) \mathcal{H}^s\left(\widetilde{A}_i\right) - (s) \int_E f(t) d\mathcal{H}^s(t) \right| < \varepsilon \tag{42}$$

holds for each partition $P = \left\{\left(\widetilde{A}_i, x_i\right)\right\}_{i=1}^{p}$ of E with $\mathcal{H}^s\left(\widetilde{A}_i\right) < \delta, i = 1, 2, \cdots, p$.

Hence

$$\sum_{i=1}^{p}\left(\sup_{\widetilde{A}_i} f(t) - \inf_{\widetilde{A}_i} f(t)\right)\mathcal{H}^s\left(\widetilde{A}_i\right) < 2\varepsilon. \tag{43}$$

Now remark that $\sum_{i=1}^{p} f(x_i)\,\mathcal{H}^s\left(\widetilde{A}_i\right)$ is the $\mathcal{H}^s$-Lebesgue integral of the $\mathcal{H}^s$-simple function $\sum_{i=1}^{p} f(x_i)\mathbf{1}_{\widetilde{A}_i}(t)$, where $\mathbf{1}_{\widetilde{A}_i}(t)$ denotes the characteristic function of $\widetilde{A}_i$; i.e., $\mathbf{1}_{\widetilde{A}_i}(t) = 1$ for $t \in \widetilde{A}_i$, and $\mathbf{1}_{\widetilde{A}_i}(t) = 0$ for $t \notin \widetilde{A}_i$.

So

$$(L)\int_{E}\inf_{\widetilde{A}_i} f(t)\cdot \mathbf{1}_{\widetilde{A}_i}(t)\, d\mathcal{H}^s(t) \le (L)\int_{\widetilde{A}_i} f(t)\, d\mathcal{H}^s(t) \le (L)\int_{E}\sup_{\widetilde{A}_i} f(t)\cdot \mathbf{1}_{\widetilde{A}_i}(t)\, d\mathcal{H}^s(t) \tag{44}$$

and

$$\begin{aligned}(L)\int_{E}\sum_{i=1}^{p}\inf_{\widetilde{A}_i} f(t)\cdot \mathbf{1}_{\widetilde{A}_i}(t)\, d\mathcal{H}^s(t) - \varepsilon \le (s)\int_{E} f(t) d\mathcal{H}^s(t) \le \\ \le (L)\int_{E}\sum_{i=1}^{p}\sup_{\widetilde{A}_i} f(t)\cdot \mathbf{1}_{\widetilde{A}_i}(t)\, d\mathcal{H}^s(t) + \varepsilon.\end{aligned} \tag{45}$$

Thus we have

$$\begin{aligned}\left|(L)\int_{E} f(t)\, d\mathcal{H}^s(t) - (s)\int_{E} f(t)\, d\mathcal{H}^s(t)\right| \le \\ \le (L)\int_{E}\sum_{i=1}^{p}\left(\sup_{\widetilde{A}_i} f(t) - \inf_{\widetilde{A}_i} f(t)\right)\mathbf{1}_{\widetilde{A}_i}(t)\, d\mathcal{H}^s(t) + 2\varepsilon < 4\varepsilon.\end{aligned} \tag{46}$$

By the arbitrariness of ε we end the proof.

Proof of Lemma 4.1.

Let c be the midpoint of $[a, b]$ and let us observe that if P_1 and P_2 are δ-fine partitions of $\widetilde{[a,c]}$ and $\widetilde{[c,b]}$, respectively, then $P = P_1 \cup P_2$ is a δ-fine partition of E. Using this observation, we proceed by contradiction.

Let us suppose that E does not have a δ-fine partition, then at least one of the intervals $\widetilde{[a,c]}$ or $\widetilde{[c,b]}$ does not have a δ-fine partition, let us say $\widetilde{[a,c]}$. Therefore $\widetilde{[a,c]}$ is not empty. Let us relabel the interval $[a, c]$ with $[a_1, b_1]$ and let us repeat indefinitely this bisection method. So, we obtain a sequence of nested intervals:
$[a,b]\supset[a_1,b_1]\supset\ldots\supset[a_n,b_n]\supset\ldots$ such that $\widetilde{[a_n,b_n]}$ is not empty. Since the length of the interval $[a_n,b_n]$ is $(b-a)/2^n$, therefore, for the Nested Intervals Property, there is a unique number $\xi\in[a,b]$ such that:

$$\bigcap_{n=0}^{\infty}[a_n,b_n] = \{\xi\}. \tag{47}$$

Let $\xi_n \in \widetilde{[a_n,b_n]}$. Therefore $|\xi_n - \xi| < |b_n - a_n| = (b-a)/2^n$. So $\lim_{n\to\infty}\xi_n = \xi$. Now since E is a closed set, $\xi \in E$.

Since $\delta(\xi) > 0$, we can find $k \in \mathbb{N}$ such that $\widetilde{[a_k, b_k]} \subseteq]\xi - \delta(\xi), \xi + \delta(\xi)[$. Therefore $\left\{([a_k, b_k]\ , \xi)\right\}$ is a δ-fine partition of $\widetilde{[a_k, b_k]}$, contrarily to our assumption.

Proof of Theorem 4.2.

By Vitali-Carathéodory Theorem, given $\varepsilon > 0$ there exist functions u and v on E that are upper and lower semicontinuos respectively such that $-\infty \le u \le f \le v \le +\infty$ and $(L)\int_E (v - u)\ d\mathcal{H}^s < \varepsilon$. Define on E a gauge δ so that

$$u(t) \le f(x) + \varepsilon \text{ and } v(t) \ge f(x) - \varepsilon, \tag{48}$$

for each $t \in E$ with $|x - t| < \delta(x)$.

Let $P = \left\{\left(\widetilde{A}_1, x_1\right), \left(\widetilde{A}_2, x_2\right), \dots \left(\widetilde{A}_p, x_p\right)\right\}$ be a δ-fine partition of E. Then, for each $i \in \{1, 2, \dots, p\}$, we have

$$(L)\int_{\widetilde{A}_i} u\ d\mathcal{H}^s \le (L)\int_{\widetilde{A}_i} f\ d\mathcal{H}^s \le (L)\int_{\widetilde{A}_i} v\ d\mathcal{H}^s. \tag{49}$$

Moreover, by $u(t) \le f(x_i) + \varepsilon$ for each $t \in \widetilde{A}_i$, it follows

$$(L)\int_{\widetilde{A}_i} (u - \varepsilon)\ d\mathcal{H}^s \le (L)\int_{\widetilde{A}_i} f(x_i)\ d\mathcal{H}^s \tag{50}$$

and therefore

$$(L)\int_{\widetilde{A}_i} u\ d\mathcal{H}^s - \varepsilon\mathcal{H}^s\left(\widetilde{A}_i\right) \le f(x_i)\mathcal{H}^s\left(\widetilde{A}_i\right). \tag{51}$$

Similarly, by $v(t) \ge f(x_i) + \varepsilon$ for each $t \in \widetilde{A}_i$, it follows

$$f(x_i)\ \mathcal{H}^s\left(\widetilde{A}_i\right) \le (L)\int_{\widetilde{A}_i} v\ d\mathcal{H}^s + \varepsilon\ \mathcal{H}^s\left(\widetilde{A}_i\right). \tag{52}$$

So, for $i = 1, 2, \dots, p$, we have

$$(L)\int_{\widetilde{A}_i} u\ d\mathcal{H}^s - \varepsilon\mathcal{H}^s\left(\widetilde{A}_i\right) \le f(x_i)\mathcal{H}^s\left(\widetilde{A}_i\right) \le (L)\int_{\widetilde{A}_i} v\ d\mathcal{H}^s + \varepsilon\ \mathcal{H}^s\left(\widetilde{A}_i\right). \tag{53}$$

Hence,

$$(L)\int_E u\ d\mathcal{H}^s - \varepsilon \le \sigma(f, P) \le (L)\int_E v\ d\mathcal{H}^s + \varepsilon, \tag{54}$$

and, by (Eq. (49)),

$$(L)\int_E u\ d\mathcal{H}^s \le (L)\int_E f\ d\mathcal{H}^s \le (L)\int_E v\ d\mathcal{H}^s. \tag{55}$$

Thus

$$\left|\sigma(f,P) - (L)\int_E f\, d\mathcal{H}^s\right| \leq (L)\int_E (v-u)\, d\mathcal{H}^s + 2\varepsilon < 3\varepsilon, \tag{56}$$

and the theorem is proved.

Appendix B

The problem of primitives is the problem of recovering a function from its derivative (i.e. the problem of whether every derivative is integrable).

In the real line the Riemann integral is inadequate to solve it, in fact the function

$$F(x) = \begin{cases} x^2 \sin(1/x^2), & x \in (0,1]; \\ 0, & x = 0; \end{cases} \tag{57}$$

is differentiable everywhere on $[0, 1]$, but its derivative

$$F'(x) = \begin{cases} 2x\sin(1/x^2) - \frac{2}{x}\cos(1/x^2), & x \in (0,1]; \\ 0, & x = 0; \end{cases} \tag{58}$$

it is not Riemann integrable since it is unbounded. Moreover, a more detailed exam reveals that F' is neither Lebesgue integrable, since F is not absolutely continuous on $[0, 1]$. Therefore the Lebesgue integral does not solve the problem of primitives. So it was natural to find an integration process for which the following theorem holds:

The Fundamental Theorem of Calculus *If* $F : [a,b] \to \mathbb{R}$ *is differentiable on* $[a,b]$, *then the function* $F'(x)$ *is integrable on* $[a,b]$ *and* $\int_a^b F'(t)dt = F(b) - F(a)$.

The first solution to this problem was given in 1912 by Denjoy, shortly followed by Perron. Both definitions are constructive. While Denjoy developed a new method of integration, called totalization, that includes the Lebesgue integral and that gets the value of the integral of a function through a transfinite process of Lebesgue integrations and limit operations, Perron used an approach that does not require the theory of measure, based on families of major and minor functions previously introduced by de la Vallée Poussin. Later Hake, Alexandroff, and Looman independently proved that the Denjoy integral and the Perron integral are equivalent (see Ref. [19]). Hence from now on, we will refer to the integral of Denjoy–Perron.

Subsequently, denoted by $\omega(F,(c,d))$ the oscillation of F on a given interval $[c,d]$ (i.e. $\omega(F,(c,d)) = \sup_{(\alpha,\beta)\subset(c,d)}|F(\beta) - F(\alpha)|$), Luzin introduced the following notions of functions AC^* and ACG^*.

Definition 6.1. A function $F : [a,b] \to \mathbb{R}$ is said to be AC^* on a set $E \subset [a,b]$ if, given $\varepsilon > 0$, there is a constant $\eta > 0$ such that

$$\sum_{i=1}^{p} \omega(F,(a_i,b_i) < \varepsilon \tag{59}$$

for each finite system $\{(a_i,b_i)\}$ of nonoverlapping intervals such that $(a_i,b_i) \cap E \neq \emptyset$.

Definition 6.2. A function $F : [a,b] \to \mathbb{R}$ is said to be ACG^* on $[a, b]$ if it is continuous and there is a decomposition $[a,b] = \cup_i E_i$ with E_i closed sets, such that F is AC^* on E_i, for each i.

Then Luzin proved the following characterization of the Denjoy–Perron integrable functions: *A function* $f : [a,b] \to \mathbb{R}$ *is Denjoy–Perron integrable on* $[a, b]$ *if and only if there exists a function* $F : [a,b] \to \mathbb{R}$ *that is* ACG^* *on* $[a, b]$ *and differentiable almost everywhere on* $[a, b]$ *such that* $F'(x) = f(x)$ *almost everywhere on* $[a, b]$.

Note that the "if" part of this characterization is often called "the descriptive" definition of the Denjoy–Perron integral.

Moreover, since the constructive definition of the Denjoy-Perron integral was not as immediate as that of the Riemann integral, only a few mathematicians were interested in working with it, so the descriptive definition became the most known definition of the Denjoy–Perron integral.

So, in spite of the general conviction that no modification of Riemann's method could possibly give such powerful results as that of the Lebesgue integral, Kurzweil in Ref. [20] and Henstock in Ref. [21] introduced independently, a generalized version of the Riemann integral that is known as the Henstock-Kurzweil integral. Such an integral solves the problem of primitives and encloses the Lebesgue integral, in the sense that if a function f is Lebesgue integrable, therefore, it is also integrable in the sense of Henstock-Kurzweil and the two integrals coincide.

The method of Henstock-Kurzweil is based on the notions of *gauge* and *partition*. Precisely, given a subinterval $[a, b]$ of the real line $\mathbb{R}$, a gauge on $[a, b]$ is, by definition, any positive function δ defined on it. A partition of $[a, b]$ is, by definition, any collection $P = \{(A_i, x_i)\}_{i=1}^{p}$ of pairwise disjoint intervals A_i and points $x_i \in A_i$ such that $[a,b] = \cup_i A_i$. Moreover, P is said to be *δ-fine* whenever $A_i \subset (x_i - \delta_i, x_i + \delta_i), i = 1, 2, \ldots, p$.

Given a gage δ on $[a, b]$, the existence of δ-fine partitions of $[a, b]$ is ensured by the following lemma.

Cousin's lemma. *For any gage* δ *on* $[a, b]$, *there exists a* δ*-fine partition of* $[a, b]$.

Given a function $f : [a,b] \to \mathbb{R}$ and a partition $P = \{(A_i, x_i)\}_{i=1}^{p}$ of $[a, b]$ the Riemann sum of f with respect to P is defined as follows

$$\sum_P f = \sum_{i=1}^{p} f(x_i)\,\mathcal{L}(A_i), \tag{60}$$

Definition 6.3. Let $f : [a,b] \to \mathbb{R}$. We say that f is Henstock–Kurzweil integrable on $[a, b]$ if there exists a number $I \in \mathbb{R}$ satisfying the following condition: to each positive ε there is a gage δ on $[a, b]$ such that

$$\left| \sum_P f - I \right| < \varepsilon, \tag{61}$$

for each δ-fine partition P of $[a, b]$.

The number I is called the Henstock–Kurzweil integral of f on $[a, b]$, and it is denoted by $(HK)\int_a^b f(x)\,dx$.

The proofs of the basic properties of the Henstock–Kurzweil integral, like the linearity and the Cauchy criterion, can be found in Ref. [10].

Moreover, the integral of Henstock-Kurzweil, even if its definition is completely different from that of Denjoy-Perron, it is equivalent to the Denjoy-Perron integral. This result was shown by several mathematicians, see Refs. [22, 23] for instance, by proving the following characterization of the Henstock-Kurzweil primitives: *The family of all Henstock-Kurzweil primitives coincides with the class of all* ACG^**-functions.*

Author details

Donatella Bongiorno
University of Palermo, Italy

*Address all correspondence to: donatella.bongiorno@unipa.it

References

[1] Falconer KJ. Fractal Geometry. Mathematical Foundations and Applications. New York: Wiley; 2003

[2] Falconer KJ. The Geometry of Fractal Sets. Cambridge: Cambridge University Press; 1986

[3] Mandelbrot BB. The Fractal Geometry of Nature. San Francisco: W. H. Freeman and Company; 1982

[4] Czachor M. Relativity of arithmetic as a fundamental symmetry of physics. Quantum Studies: Mathematics and Foundations. 2016;**3**:123-133

[5] De Guzman M, Martin MA, Reyes M. On the derivation of fractal functions. In: Proc. 1st IFIT Conference on Fractals in the Fundamental and Applied Sciences. Portugal (North-Holland): Lisboa; 1991. pp. 169-182

[6] Jiang H, Su W. Some fundamental results of calculus on fractal sets. Communications in Nonlinear Science and Numerical Simulation. 1998;**3**(1): 22-26

[7] Parvate A, Gangal AD. Calculus on fractal subsets of real line - I: Formulation. Fractals. 2009;**17**(1): 53-81

[8] Bongiorno D, Corrao G. The Henstock-Kurzweil-Stieltjes type integral for real functions on a fractal subset of the real line. Bollettino di Matematica pura e applicata vol. IV. 2011;**4**:1-19

[9] Bongiorno D, Corrao G. On the fundamental theorem of calculus for fractal sets. Fractals. 2015;**23**(2):10. DOI: 101142/50218348X15500085

[10] Gordon RA. The Integrals of Lebesgue, Denjoy, Perron, and Henstock. Graduate Studies in Mathematics (Volume 4). Providence, Rhode Island: American Mathematical Society; 1994

[11] Golmankhaneh AK, Baleanu D. Fractal calculus involving gauge function. Communications in Nonlinear Science and Numerical Simulation. 2016; **37**:125-130

[12] Golmankhaneh AK, Baleanu D. New derivatives on the fractal subset of real-line. Entropy. 2016;**18**(2):1-13. DOI: 10.3390/e18020001

[13] O'Malley RJ. First return path derivatives. Proceedings of the American Mathematical Society. 1992;**116**:73-77

[14] Freiling C. The equivalence of universal and ordinary first-return differentiation. Real Analysis Exchange. 2000/2001;**26**:5-16

[15] Darji UB, Evans MJ. A first return examination of the Lebesgue integral. Real Analysis Exchange. 2001/2002; **27**(2):573-582

[16] Mattila P. Geometry of Sets and Measures in Euclidean Spaces. Fractal and Rectificability. Cambridge: Cambridge University Press; 1995

[17] Bongiorno D, Corrao G. An integral on a complete metric measure space. Real Analysis Exchange. 2015;**40**(1):157-178

[18] Bongiorno D. Derivatives not first return integrable on a fractal set. Ricerche di Matematica. 2018;**67**(2):597-604

[19] Saks S. Theory of the Integral. New York: Dover; 1964

[20] Kurzweil J. Generalized ordinary differential equations and continuous

dependence on a parameter. Czechoslovak Mathematical Journal. 1957;7:418-446

[21] Henstock R. Definition of Riemann type of the variational integral. Proceedings of the London Mathematical Society. 1961;**11**(3):402-418

[22] Kubota Y. A direct proof that the *RC*-integral is equivalent to the D^*-integral. Proceedings of the American Mathematical Society. 1980;**80**:293-296

[23] Gordon RA. Equivalence of the generalized Riemann and restricted Denjoy integrals. Real Analalysis Exchange. 1986/1987;**12**(2):551-574

Chapter 4

Generalised Z-Entropy (Gze) and Fractal Dimensions (FDs)

Ashiq Hussain Bhat and Ismail A. Mageed

Abstract

In this chapter, we explore the relationship between FD, a statistical measure that gauges the complexity of a given pattern embedded in a certain set of spatial dimensions, and GZe. To interpret the behaviour of the derived (Gze) fractal dimension matching to its parameters, numerical tests are conducted. The broadest generalisation in the literature is reported in this chapter. More potentially, chapter explores the relationship between fractal dimension, which measures the complexity of a pattern in space, and Gze, the ultimate entropy in the literature. The chapter also integrates information theory with fractal geometry, and numerical experiments are conducted to interpret the behaviour of the derived Gze fractal dimension. This work is considered the ultimate generalisation in the literature.

Keywords: shannon's entropy, FD (fractal dimension), fractal geometry, GZe, scaling factor

1. Introduction

Claude Shannon defined the entropy H for a discrete random variable X in 1948 [1]; as given by

$$H(X) = \sum_i p(x_i) I(x_i) = -\sum_i p(x_i) \ln(p(x_i)) \tag{1}$$

Here $p(x_i)$ serves as the ith-event probability. There are two mathematical concepts: entropy and fractal dimension. Entropy measures the information in a distribution, while FD measures a spatial pattern's complexity in space and how it fills the available space. The idea of using fractional calculus to measure FDs has been around for a long time, but it became popular thanks to Benoit Mandelbrot's work on fractional dimensions in 1967 [2].

A coastline's FD [2] can be measured more fundamentally. The idea came from a prior work by Lewis Fry Richardson, who noted that the measured length of a coastal line varies depending on the length of the measuring stick. The number of sticks used to measure the shoreline and the scale of the stick used to determine FD.

The complexity of patterns [3]; in different spatial dimensions can be investigated using a statistical index called the fractal dimension. In this framework, several formal mathematical definitions of fractal dimension exist. There are two formulas that use the number of sticks used to cover the coastline (N), the scaling factor (ϵ), and the fractal dimension (D) to determine the fractal dimension of a given pattern.

$$N \propto \epsilon^{-D} \tag{2}$$

$$\ln N = -D = \frac{\ln N}{\ln\ \epsilon} \tag{3}$$

Let us see this example. To replicate what Richardson examined, we use Google Earth satellite pictures and GIMP (the GNU Image Manipulation Programme) to create a map and rigid sticks [4]; **Figure 1** depicts the similar approach for a section of the Grand Canyon. The Google Earth ruler tool is used to determine the reference length. The rulers for 6 km, 3 km, and 1 km can be found in the left-upper panel. We use a 6 km reference length to calculate the fractal dimension. We can see in the left-lower panel that we need roughly 13 stiff sticks, each one-half the reference length long, to follow the rim of this section of the canyon. The table shows the fractal dimension for various numbers of sticks at different lengths.

The reference length of 6 km is established using Google Earth's ruler tool by employing rigid sticks in chase of the Canyon's rim. The number of sticks needed (N) depends on their length, with shorter sticks requiring more images. The fractal dimension of the rim is given in **Table 1**.

Table 1 describes a method for experimentally determining the fractal dimension of a canyon's rim. This explains the concept of discrete uniform distribution in probability, using the example of throwing a die with six faces and the probability of obtaining a given score. In **Figure 2** below, the significant impact of N on ϵ is illustrated. More interestingly, **Figure 3** provides strong supporting evidence of the impact of FD on ε (the scaling factor). Clearly, by looking at **Figure 2**, we can see that the scaling factor decreases with a very heavy tailed trend by the increase of N, whereas, in **Figure 3**, the portrayed data shows that both ε (the scaling factor) and FD are decreasing at the same time.

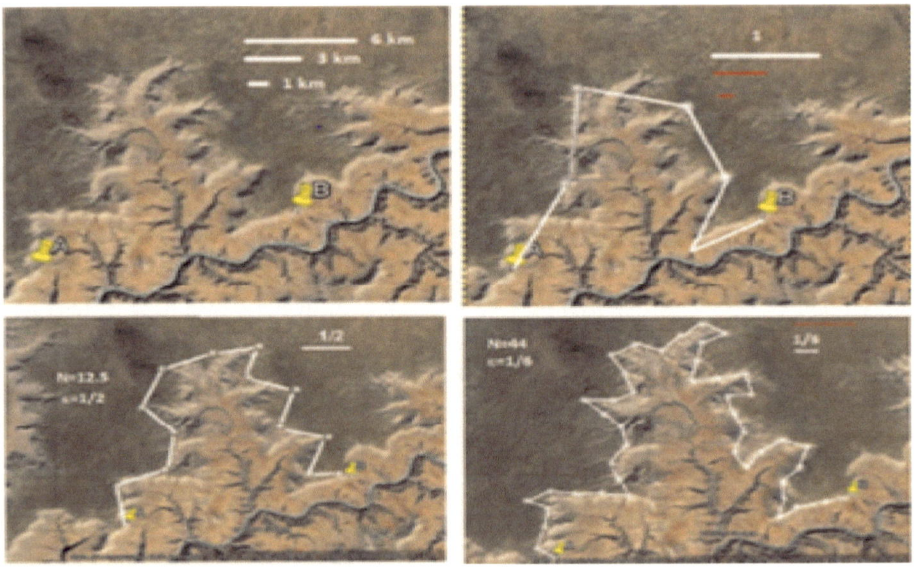

Figure 1.
Google earth satellite images and GNU image manipulation program of a part of grand canyon, Arizona.

N	ϵ	D
13	½	3.70
44	⅙	2.11
119	1⁄12	1.92
405	1⁄30	1.72
871	1⁄60	1.65

Table 1.
Fractal dimension of a canyon's rim.

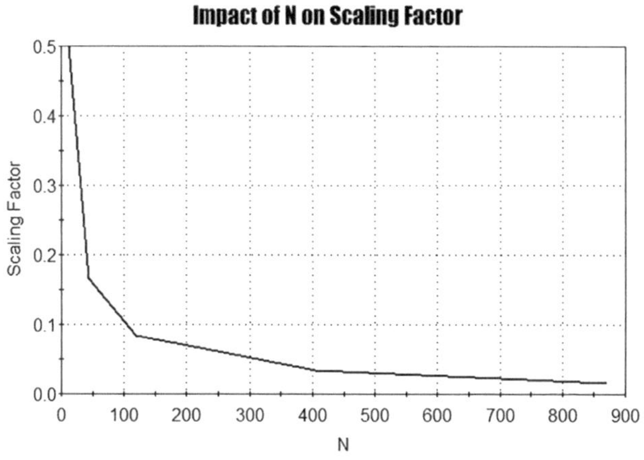

Figure 2.
An illustrative data portrait of how N impacts ϵ.

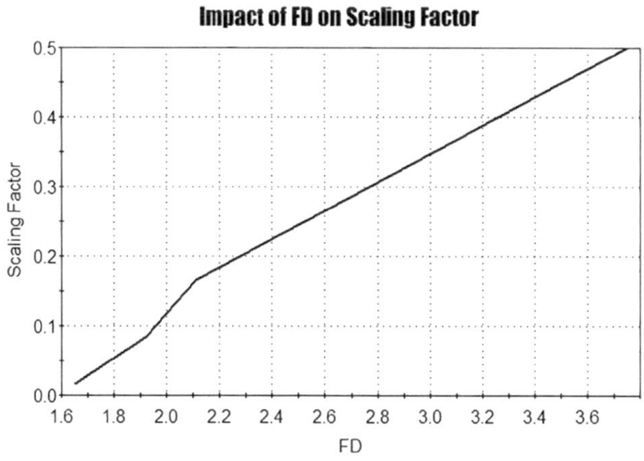

Figure 3.
An illustrative data portrait of how FD impacts ϵ.

The current chapter road map is: The existent literature on entropic derivation of fractal dimension is overviewed in Section 2. Section 3 mainly deals with the derivation of the new results and provides numerical portrait which clearly supports the strong evidence of the significant impact of the GZe's parameters on the behaviour of Gze's FD. Section 4 is devoted to conclusion and future work.

2. Materials and methods

Various generalised entropy measures have been developed by various authors in the last few decades. Bhat and Baig [5–11]; also developed various generalised entropy measures to the literature of information theory. For the equiprobable distribution's assumption, we have used the definition of Shannon entropy [1]; Rényi entropy [12, 13]; Tsallis entropy [14]; and Kaniadakis entropy [15]; to obtain our revolutionary derivations. The Shannonian FD [4]; is given by:

$$D_s = \lim_{\epsilon \to 0} \frac{\ln N}{\frac{1}{\epsilon}} \tag{4}$$

The Rényian FD for $q \in (0.5,1)$, is defined in the following manner [13].

$$D_R = \lim_{\epsilon \to 0} \frac{\ln N}{\frac{1}{\epsilon}} \tag{5}$$

The Tsallisian FD for $q \in (0.5,1)$, is defined in the following manner [4].

$$D_T = \lim_{\epsilon \to 0} \frac{\frac{1}{1-q}\left(N^{1-q} - 1\right)}{\frac{1}{\epsilon}}, D_T > 0 \tag{6}$$

The Kaniadakisian FD [8]; with k to serve as the entropic index is determined by:

$$D_K = \lim_{\epsilon \to 0} \frac{\frac{1}{2k}\left(N^k - N^{-k}\right)}{\frac{1}{\epsilon}} \tag{7}$$

The behaviour of these generalised dimensions when their indices are varied can be seen. The case of the Koch snowflake $N = 4$ and $\epsilon = \frac{1}{3}$, was proposed in **Figure 3** to determine the corresponding fractal dimension to each entropy. It is observed from **Figure 3**, that Tsallis generalised statistics seem to be the natural frame for studying fractal systems.

3. Results and discussion

The proposed Generalised Z-Entropy (Gze), [16]; is a non-extensive entropy functional defined by:

$$H_{q,a,b,Z}(p) = Z_{a,b} = \frac{c}{(1-q)(a-b)}\left[\left(\sum_n p_{q,Z}(n)^q\right)^a - \left(\sum_n p_{q,Z}(n)^q\right)^b\right] \tag{8}$$

where c is a positive constant, $1 > q > 0.5, a > 0, b \in R \; or \; b > 0, a \in R \; with \; a \neq b$.

The following proposition is of great importance as it solidifies our choice of using the proposed GZE in our study.

Preposition 3.1: (c.f., P. Tempesta, [16])

1. The $Z_{0,0}$ is the Rényian entropy.

2. The $Z_{1,0}$ is the Tsallisian entropy.

3. The $Z_{k,-k}$ is the $Z_{k,q}$ = k-entropy, namely Kaniadakis entropy.

4. The $Z_{a,0}$ is Sharma–Mittal entropy [17]

$$Z_{0,0}, \mathrm{q} \to 1 \text{ is Shannon's entropy.}$$

Consequently, $Z_{a,b}$ – entropy is ultimately most available entropies in the literature.

Theorem 3.2: In the case of equiprobable distribution, the GZE fractal dimension, $D_{a,b}$ is devised by.

$$D_{Z_{a,b}} = \lim_{\epsilon \to 0} \frac{\frac{1}{(1-q)(a-b)}\left(N^{(1-q)a} - N^{(1-q)b}\right)}{\frac{1}{\epsilon}} \tag{9}$$

Provided that, $1 > q > 0.5, a > 1, b \in R \; or \; b > 0, a \in R \; with \; a \neq b.$

Proof:

It can be seen that:

$$D_{Z_{a,b}} = \frac{1}{(1-q)(a-b)} \lim_{\epsilon \to 0} \frac{\left[\left(\sum_n p_{q,Z}(n)^q\right)^a - \left(\sum_n p_{q,Z}(n)^q\right)^b\right]}{\frac{1}{\epsilon}}$$

$$= \frac{1}{(1-q)(a-b)} \lim_{\epsilon \to 0} \frac{\left(\frac{1}{N}\right)^{aq}(N)^a - \left(\frac{1}{N}\right)^{bq}(N)^b}{\frac{1}{\epsilon}}$$

$$= \frac{1}{(1-q)(a-b)} \lim_{\epsilon \to 0} \frac{\left(N^{(1-q)q} - N^{(1-q)b}\right)}{\frac{1}{\epsilon}}$$

as claimed (c.f., (9)).

Corollary 3.3: The GZE fractal dimension, $D_{Z_{a,b}}$ satisfies the following:

i. $\lim_{a \to 0, b \to 0} D_{Z_{a,b}} = D_R$

ii. $\lim_{q \to 1} \left(\lim_{a \to 0, b \to 0} D_{Z_{a,b}}\right) = D_s$

iii. $\lim_{a \to 1, b \to 0} D_{Z_{a,b}} = D_T$

iv. $\lim_{a \to k, b \to -k} D_{Z_{a,b}} = D_K$

v. $\lim_{b \to 0} D_{Z_{a,b}} = D_{Sharma-Mittal}$

Proof:

It could be verified that $D_{Z_{a,b}}$ (c.f., (9)) satisfies the following:

$$\lim_{a\to 0, b\to 0} D_{Z_{a,b}} = \lim_{\epsilon\to 0} \frac{\lim_{a\to 0} \frac{1}{(1-q)(a)}\left(N^{(1-q)a}-1\right)}{\frac{1}{\epsilon}}$$
$$= \lim_{\epsilon\to 0} \frac{\lim_{a\to 0,} \frac{1}{(1-q)(1)}\left((1-q)N^{(1-q)a}\ln N\right)}{\frac{1}{\epsilon}} \quad \left(\text{L'Hopital's Theorem of Limits}\right)$$
$$= \lim_{\epsilon\to 0} \frac{\ln N}{\frac{1}{\epsilon}} = D_R \ (\text{c.f.}, (5)) \tag{10}$$

Taking the limit of both sides of (10) as $q \to 1$, we have

$$\lim_{q\to 1}\left(\lim_{a\to 0, b\to 0} D_{Z_{a,b}}\right) = \lim_{q\to 1}\left(\lim_{\epsilon\to 0}\frac{\ln N}{\frac{1}{\epsilon}}\right) = \lim_{\epsilon\to 0}\frac{\ln N}{\frac{1}{\epsilon}} = D_s \ (\text{c.f.}, (4)) \tag{11}$$

$$\lim_{a\to 1, b\to 0} D_{Z_{a,b}} = \lim_{\epsilon\to 0} \frac{\lim_{a\to 1} \frac{1}{(1-q)(a)}\left(N^{(1-q)a}-1\right)}{\frac{1}{\epsilon}} = \lim_{\epsilon\to 0} \frac{\frac{1}{(1-q)}\left(N^{(1-q)}-1\right)}{\frac{1}{\epsilon}} = D_T \ (\text{c.f.}, (6)) \tag{12}$$

$$\lim_{a\to k, b\to -k} D_{Z_{a,b}} = \lim_{\epsilon\to 0} \frac{\lim_{a\to k, b\to -k} \frac{1}{(1-q)(a-b)}\left(N^{(1-q)a}-N^{(1-q)b}\right)}{\frac{1}{\epsilon}}$$
$$= \lim_{\epsilon\to 0} \frac{\frac{1}{(1-q)(2k)}\left(N^{(1-q)k}-N^{-(1-q)k}\right)}{\frac{1}{\epsilon}}$$

Define $(1-q)k = \lambda_k, c = \frac{1}{1-q}, q \in (0.5,1)$
Hence, it follows that.

$$\lim_{a\to k, b\to -k} D_{Z_{a,b}} = \lim_{\epsilon\to 0} \frac{\frac{c}{2\lambda_k}\left(N^{\lambda_k}-N^{-\lambda_k}\right)}{\frac{1}{\epsilon}} = D_K \ (\text{c.f.}, (7)) \tag{13}$$

$$\lim_{b\to 0} D_{Z_{a,b}} = \lim_{\epsilon\to 0} \frac{\lim_{b\to 0} \frac{1}{(1-q)(a-b)}\left(N^{(1-q)a}-N^{(1-q)b}\right)}{\frac{1}{\epsilon}}$$
$$= \lim_{\epsilon\to 0} \frac{\frac{1}{(1-q)(a)}\left(N^{(1-q)a}-1\right)}{\frac{1}{\epsilon}} = D_{Sharma,Mittal} \tag{14}$$

We have determined that:

$$D_{Z_{a,b}} = \lim_{\epsilon\to 0} \frac{\frac{1}{(1-q)(a-b)}\left(N^{(1-q)a}-N^{(1-q)b}\right)}{\frac{1}{\epsilon}} \ (\text{c.f.}, (9)).$$

Following the Koch snowflake ($N = 4$ *and* $\epsilon = \frac{1}{3}$), we have

$$D_{Z_{a,b}} = \frac{3}{(1-q)(a-b)}\left(4^{(1-q)a} - 4^{(1-q)b}\right) \tag{15}$$

It can be seen that:

$$D_{Z_{2,-1}} = \frac{1}{(1-q)}\left(4^{2(1-q)} - 4^{(q-1)}\right) \tag{16}$$

And

$$D_{Z_{-2,1}} = -\frac{1}{(1-q)}\left(4^{-2(1-q)} - 4^{(1-q)}\right) \tag{17}$$

Moreover, it holds that:

$$\lim_{q\to 1} D_{Z_{2,-1}}(q) = \lim_{q\to 1} \frac{1}{(1-q)}\left(4^{2(1-q)} - 4^{(q-1)}\right)$$
$$= \ln 4 \lim_{q\to 1}\left(2\left(4^{2(1-q)}\right) + 4^{(q-1)}\right) = 3\ln 4 = 4.158883083 \tag{18}$$

Also, we have

$$\lim_{q\to 1} D_{Z_{-2,1}}(q) = \lim_{q\to 1} \frac{1}{(1-q)}\left(4^{-2(1-q)} - 4^{(1-q)}\right)$$
$$= \ln 4 \lim_{q\to 1}\left(2\left(4^{-2(1-q)}\right) + 4^{(1-q)}\right) = 3\ln 4 = 4.158883083 \tag{19}$$

Thus, we see that

$$\lim_{q\to 1} D_{Z_{2,-1}}(q) = \lim_{q\to 1} D_{Z_{-2,1}}(q) = 4.158883083$$

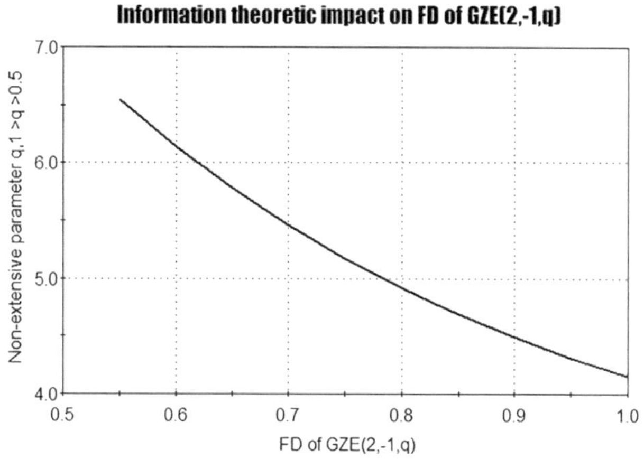

Figure 4.
Decreasability of $D_{Z_{2,-1}}(q)$ *against the non-extensive information theoretic parameter* $q, 1 > q > 0.5$.

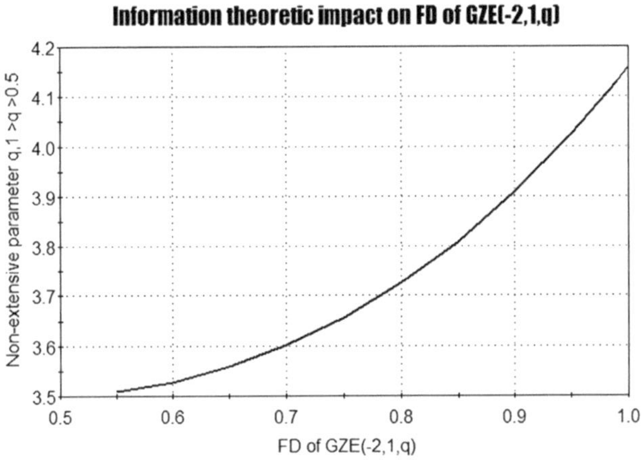

Figure 5.
Increasability of $D_{Z_{-2,1}}(q)$ *against the non-extensive information theoretic parameter* q*,* $1 > q > 0.5$.

It is observed from **Figures 4** and **5**, that $D_{Z_{2,-1}}(q)$ decreases for $q \in (0.5,1)$, while $D_{Z_{-2,1}}(q)$ increases for $q \in (0.5,1)$.This provides a strong evidence to support the significant impact of the non-extensive information theoretic parameter q, $(1 > q > 0.55)$ on the overall behaviour of both $D_{Z_{2,-1}}(q)$ and $D_{Z_{-2,1}}(q)$.

4. Conclusion and future work

This study explores the relationship between Gze and FD, the latter measures the complexity of a pattern. The study uses numerical experiments to understand how Gze behaves based on its parameters. The study is considered a significant step in bringing together information theory and fractal geometry. Future work involves finding the fractal dimension of available entropies in the literature to draw a detailed comparison between these derived fractal dimensions, which will open new grounds towards Information Theoretic Fractal Geometry (ITFG).

Author details

Ashiq Hussain Bhat[1*] and Ismail A. Mageed[2]

1 Department of Bio-Statistics, Indian Council of Medical Research-RMRIMS, Patna, India

2 Department of Computer Science, University of Bradford, United Kingdom

*Address all correspondence to: ashiqhb14@gmail.com

References

[1] Shannon CE. A mathematical theory of communication. Bell System Technical Journal. 1948;**27**:379-423

[2] Mandelbrot BB. The Fractal Geometry of Nature. New York: W.H. Freeman; 1983

[3] Harte D. Multifractals. London: Chapman and Hall; 2001

[4] Sparavigna AC. Entropies and Fractal Dimensions. 2016. PHILICA.COM. Article number 559

[5] Bhat AH, Baig MAK. Characterization of new two parametric generalized useful information measure. Journal of Information Science Theory and Practice. 2016;**4**(4):64-74

[6] Bhat AH, Baig MAK. Noiseless coding theorems on new generalized useful information measure of order alpha and type beta. Asian Journal of Fuzzy and Applied Mathematics. 2016; **4**(6):73-85

[7] Bhat AH, Baig MAK. New generalized measure of entropy of order and type and its coding theorems. International Journal of Information Science and System. 2016;**5**(1):1-7

[8] Bhat AH, Baig MAK. Some coding theorems on generalized Reynis entropy of order alpha and type beta. International Journal of Applied Mathematics and Information Sciences Letters. 2017a;**5**(1):13-19

[9] Bhat AH, Baig MAK. New generalized entropy measure and its corresponding code-word length and their characterizations. International Journal of Advance Research in Science and Engineering. 2017;**6**(1): 863-873

[10] Bhat AH, Baig MAK. Coding theorems on new additive information measure of order alpha. Pakistan Journal of Statistics. 2018;**34**(2):137-146

[11] Bhat AH, Dar JA, Baig MAK. Two parametric new generalized average code-word length and its bounds in terms of new generalized inaccuracy measure and their characterization. Pakistan Journal of Statistics. 2018;**34**(2): 147-162

[12] Renyi A. On measures of information and entropy. Proceedings of the Fourth Berkeley Symposium on Mathematical Statistics and Probability. 1961;**4**:547-562

[13] Ott E. Attractor dimensions. Scholarpedia. 2008;**3**(3):2110

[14] Tsallis C. Possible generalization of Boltzmann-Gibbs statistics. Journal of Statistical Physics. 1988;**52**(1):479-487

[15] Kaniadakis G. Statistical mechanics in the context of special relativity. Physical Review E. 2002;**66**(5):056-125

[16] Tempesta P. Formal groups and *Z*-entropies. Proceedings: Mathematical, Physical and Engineering Sciences. Nov 2016;**472**(2195):20160143. DOI: 10.1098/rspa.2016.0143. PMID: 27956871; PMCID: PMC5134302

[17] Sharma BD, Mittal DP. New nonadditive measures of entropy for discrete probability distributions. Journal of Mathematical Sciences. 1975; **10**:28-40

Chapter 5

Fractal Analysis of Façades of Historical Public Buildings with Box Count Method: The Case of Afyonkarahisar

Gamze Çoban and Şerife Ebru Okuyucu

Abstract

This chapter aims to analyze the fractal geometry of the façade plane of 3 (three) historical public buildings in Afyonkarahisar city center by box counting method. Within the scope of the study, 3 (three) historical public buildings belonging to the First National Architecture period were selected. The examples in this study were selected from the entrance façades of historical public buildings in the city of Afyonkarahisar. According to the historical public building façades defined, the dimensions of the façades remained in their original state, the façade reliefs of 3 (three) sample buildings were taken and their two-dimensional front views were drawn in vector, and the fractal fiction between each other was analyzed using the "box counting method". Fractal analysis was performed by box counting method in the context of occupancy-vacancy ratios formed by building elements such as windows, doors, jambs, floor erasures, buttresses, eaves that make up the façade structure. As a result of the study, how the façade features of the First National Architecture period corresponded to the façade constructions of the selected public buildings were interpreted based on numerical data obtained from fractal analysis.

Keywords: fractal analysis, box counting method, historical public building façades, Afyonkarahisar, fractal geometry

1. Introduction

The First National Architecture movement covers a period between 1908 and 1930 that emerged as an anti-Westernization movement in Turkey. On the basis of the First National Architecture movement, there is the idea of giving Turkish identity to buildings by adapting Turkish Seljuk and Turkish Ottoman architectural forms to the new functions of contemporary architecture. In this context, an eclectic architectural approach called "national style" was seen in the buildings built during this period. In this period, especially the façade designs of public buildings were given great importance in the context of image and Ottoman architectural elements (dome, arch, portico, crown door, etc.) manifested themselves in the form of the use of ornamental motifs and applied in new layouts on the façades. In this context, in this study, the mathematical expressions of the façade designs of public buildings produced in the First National Architecture period were calculated by fractal analysis method.

There are very few quantitative methods that can measure the thousands of architectural lines (such as windows, doors, pilasters, buttresses, jambs, floor moldings, eaves lines) on the façade of a historical building of a certain period. One of the methods used for this purpose is fractal analysis. The method of fractal analysis is used to calculate two-dimensional linear constructions. Fractal geometry is a geometric form formed by the combination of complex, repetitive and continuous fictions [1].

Within the scope of the study, three public buildings with different functions were selected from the First National Architecture period in Afyonkarahisar: "Zafer Museum", "Rıza Çerçel Cultural Center" and "Police Station Building". The construction dates of these three public buildings are close to each other and it is seen that the façade designs are compatible with the features of the First National Architectural Period. The architectural style and eclectic approach in the façade designs of public buildings of this period are expressed in complex lines on the façade. In order to analyze the façade structure of these structures, a mathematical expression was used and façade readings were made by evaluating the occupancy-gap ratios with numerical data based on fractal geometry. While the façade readings of the public buildings were made, the occupancy- vacancy ratios were evaluated in the context of the frames formed by the building elements such as windows, doors, jambs, floor moldings, eaves and buttresses on the façade. In order to perform fractal analysis on the façades of public buildings, first of all, façade survey drawings of public buildings were prepared. In the study, the structures were analyzed using façade drawings. The analysis was performed using the "box counting method" within the mathematical concept of "fractal geometry". The calculated fractal dimension value was obtained for each structure. Box counting method is a technique used to calculate and read the complexity of architectural façades and plans. This method takes into account the richness of detail and reconstruction of various textures with two-dimensional complex structures that cannot be calculated by other methods.

In this context, fractal analyses made on the façades of three different public structures belonging to the First National Architectural Period, "Zafer Museum", "Rıza Çerçel Cultural Center" and "Police Station Building", were converted into numerical data by box counting method. In line with the numerical data obtained, a comparison was made between the façade structures of the public buildings of the "Zafer Museum", "Rıza Çerçel Cultural Center" and "Police Station Building" and the façade features of the First National Architectural Period. After all, it was seen that the numerical expressions corresponding to the mobility and complexity in the façade constructions of these structures coincided with the data obtained from the fractal analysis using the box counting method.

2. Features of the first national architectural period

After the declaration of the Second Constitutional Monarchy in 1908, architecture emerged as one of the key domains that significantly shaped the conditions and structure of the period, alongside the political, social, economic and cultural initiatives of the Union and Progress Party. In this period, it is seen that many thinkers, such as Ziya Gökalp pioneered the way of awakening national consciousness, reviving it and adopting concepts such as returning to the essence. In particular, the ideas developed by Gökalp in the fields of economy, politics, philosophy, law, religion and language became widespread over time and this situation allowed to create an environment suitable for developments in the field of architecture [2].

The First National Architecture period, which emerged as a result of the Turkism Movement influencing architecture as well as in every field and which existed between 1908 and 1930, was a process in which there was a reaction against Westernization in architecture and foreign architects [3]. Architects such as Kemalettin Bey, Vedat Tek, Muzaffer Bey, Arif Hikmet Koyunoğlu, Alexander Valaury, Guilio Mongeri, who pioneered this movement, examined the elements of past architecture and created the works of the First National Architectural Period, which was the Neoclassical style of the period [2, 4–6]. A national architectural movement was tried to be established by placing building elements representing the classical period Ottoman and Seljuk architecture on the Neorenaissance building masses of Western origin [2]. For this reason, it is possible to see the influences of the Bat culture at the root of the National Architecture that began to develop in the twentieth century and became stronger after 1908. In the buildings of the First National Architectural Period, the influences of foreign architects and Western artists can be seen very clearly. It is possible to list the general characteristics of the works of this period as follows:

- There is a symmetrical and axial understanding of mass arrangement and plan schemes that are based on the examples.

- The façades are symmetrical and majestic. The use of Ottoman and Seljuk motifs in the buildings built in this period, especially on the exteriors, supported the formation of national consciousness [4].

- The schemes used are parallel to European Neoclassicism [7] in terms of the dimensions and proportions of the spaces, the façade arrangements and the composition rules. Apart from religiously functional structures, all types of buildings, especially their facades, are divided into three sections with continuous stone belts, in accordance with the rules of Neorenaissance architecture. Each section is organized in its own right as a whole. On each floor, different window forms are applied. The structures are given a certain rhythmic appearance in vertical order.

- The national body has tried to be used in non-structural elements and decoration with elements selected from the classical Ottoman religious architecture in the fifteenth and sixteenth centuries. However, in the surface arrangement methods, the understanding of form of Western architecture is also dominant [8]. In this arrangement, which formed the basis of nineteenth-century European architecture, is stated that instead of the Classical Greek and Roman styles in Europe, building elements such as flattened or pointed arches, column capitals with beads or mukarnas, moldings decorated with geometric reliefs, tile panels, wide eaves with ornamented lower surfaces and false domes were used [5].

- Another decisive feature of buildings built in the First National Architectural style is the tower, which began with Art Nouveau long before the use of the tower in Turkey. Art Nouveau [9] is a style that is not included in the classical tradition of Europe and is fed by many sources, so it came to Turkey in forms originating from North Africa, East and Far Abroad. The First National Architects, who had to adapt the knowledge and values they acquired from Europe or through their education to the conditions of Turkey, also manifested the emergence of Turkish Neoclassicism by incorporating these styles [10].

- The Seljuk and Ottoman architectural elements used by Turkish architects during the First National Architectural Period were not always in harmony with the functions of the buildings they designed. With these works, the architects tried to respond to the functional needs of the society as well as the integrative symbolic needs that emphasize the identity of the nation. In this understanding of architecture, they combined architectural and ornamental elements taken from Seljuk and Ottoman architecture with the design principles and construction techniques of Europe. In accordance with Turkish identity, the concept of "nationality" replaces religion [11].

According to Ünsal [12], the selective style adopted during the First National Architecture period, which encompassed national feeling and thought, served as a fitting attire for the new buildings that defined the city. This style was particularly prominent in the construction of banks, post offices, educational buildings, hotels, other public structures. Ünsal [12] stated that the plan was not taken into consideration in this period, that the façades were prioritized and if the façade was good, the survey of the plan was carried out.

Despite the "Turkish" architectural features sought in the buildings of the First National Architectural Period, a nationalist approach was not exhibited in terms of materials and mostly materials brought from the West were used. However, although the materials were imported from the West, it was emphasized by the architects that the ways of using these materials should be compatible with the architectural understanding of the period. During this period, the main idea was to combine decorative elements taken from classical Ottoman architecture, especially Ottoman domes in the form of half domes, wide roof consoles with supporting elbows, pointed arches and new construction techniques such as reinforced concrete with tile decoration. A kind of Turkish eclecticism has been created by the domes used, oval arched windows, hedgehog eaves, tile panels, ornate front façades, rhythmic window layouts on the façades, compartment-like interior designs consisting of spaces placed on both sides of a corridor, hipped roof application, separation of façades with moldings, animation with plasters. Considering the financial conditions and urban scale of the period, most, if not all, of them were large buildings and were built on multiple floors. A structure similar to European Neoclassicism is observed in terms of the plans and schemes used, the measurements and proportions in masses and spaces, the composition rules and multi-storey construction techniques [13].

It is possible to see the first examples of architectural structures of the First National Architectural Movement starting from 1900 until the first years of the establishment of the Republic. During the effort to create a new nation, construction initially focused on public buildings (such as banks, hospitals, palaces, schools), then built apartment buildings and business khans [14].

3. Fractal analysis method in architecture

To date, architecture has been analyzed by many different methods. In studies focusing on the relationship between architecture and the user, it is seen that methods such as survey, simulation techniques, interviews, physical traces, behavior observation are used. Today, these methods, which are not based on numerical data and are intuitive, have been replaced by methods that reveal more systematic and numerical data. One of these methods is the Fractal Theory.

Fractal theory is the analysis method of this study. It was first introduced in 1975 by the mathematician B. Mandelbrot for the description of forms that repeat each other at different scales. B. Mandelbrot published an article in the journal "Science" entitled "How long is the length of the British coastline", stating that there is an inversely proportional relationship in which the length of the coastline varies depending on the length of the measurement used to measure the coastline, and that the length of the coastline lengthens as the length of the instrument decreases [1, 15]. Derived from the Latin word "fractus", which means broken, fragmented, and the English word "fraction", which means part, division, fractal fictions describe a universe consisting of indented protruding, broken, bent, interlocking shapes that cannot be round or straight lines [16, 17]. It is stated that even when zooming in on fractals, which are very complex and have unlimited detail, the detail of the whole shape can be seen, and similar shapes will emerge regardless of how close or distant the shape is. When defined as a fractal function, "x, f(x), f(f(x)), ... ", that is, it can be repeated forever [18].

Although it was first named in 1975, it was founded in the twentieth century. At the beginning of the century, many mathematicians used forms that showed fractal properties. These forms, which are also the source of Mandelbrot's work, have been called monsters because of their irregular, complex structure. The Cantor Set, introduced by George Cantor and forming the basis of set theory in mathematics, underpinned many fractal fictions and has been called classical fractals (**Figure 1**) [19].

It is stated that a structure must meet certain conditions to be called fractal and the conditions are;

- having a perfect structure,
- being too indented to be explained by Euclidean geometry,
- self-similarity,
- having a fractional size,
- occurrence as a result of a simple rule of repetition [20].

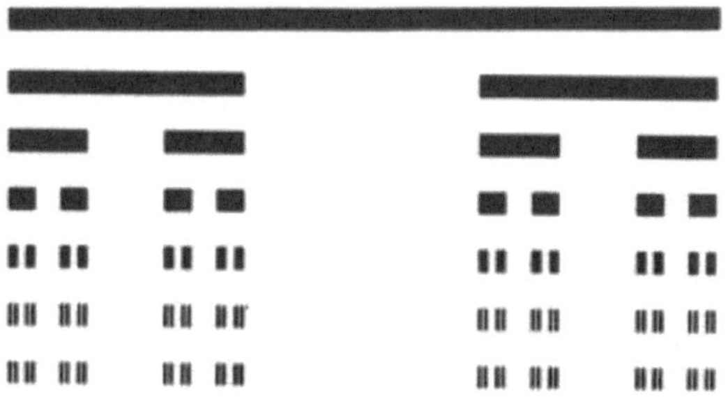

Figure 1.
Cantor set [19].

Fractals, which exhibit self-similarity and are found in various forms, can be classified in multiple ways based on different criteria. According to Mandelbrot, fractals are classified into man-made artificial fractals and natural fractals [18].

Natural fractals exist in nature in their own form and can be found in a maximum of 3 or 4 cycles. Fractals in nature lose detail in their structure during any cycle. At the core of natural fractals, they have the property of their own resemblance, and these formations can be found in the structure, in the scale of the tissue. Clouds, stars, plants, mountains, boundary lines, nerves in the human body, age rings that appear when the trunk of a tree is cut down can be cited as examples of natural fractals (**Figure 2**) [21, 22].

Artificial fractals are produced by humans, which can be found in an infinite number of cycles. These fractals created in the mathematical environment are also called generative algorithms, which are formed within the framework of certain geometric rules. All the dividing parts of the artificial fractals have the same detail and are exactly similar to the whole fractal. An infinite number of self-repeating geometric shapes form ideal fractal patterns, and they appear only in mathematical theories or digital graphs. The Koch curve, the Sierpinski triangle, the Minkowski curve, the Barnsley fern are cited as examples [23]. When the Koch curve example is examined, it is seen that the first cycle is completed by dividing the structure with a line, dividing the line into three equal parts and transforming the middle part into an equilateral triangle and applying the same rule to each line. This process is repeated to produce a snowflake like shape (**Figure 3**) [23].

In fractal geometry, the value obtained as a result of measuring the complexity of a fractal is defined as the fractal dimension, and the high fractal dimension value indicates that the complexity of the surface is high. In the context of fractal analysis, when the obtained fractal value falls closer to 1, it is generally considered an indicator of simplicity and less complexity. On the other hand, when the value approaches 2, it suggests increased complexity [21]. Different methods have been developed for calculating the fractal dimension, which was introduced in the 1970s. Self-similarity

Figure 2.
Examples of natural fractals [24–27].

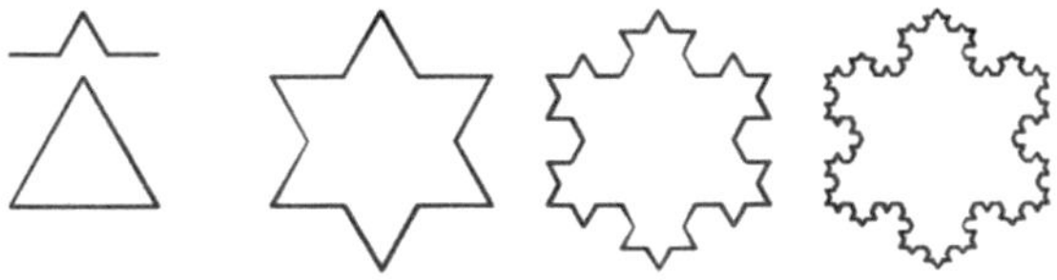

Figure 3.
Koch curve [27].

dimension, measure dimension and box counting dimension are some of these methods [21]. In self-similarity dimension calculations, it is stated that a curve similar to itself should consist of repeating parts, and that the self-similarity method is not suitable for calculation because of the small number of repetitive structures in two-dimensional complex constructions. The measure dimension method cannot be used in two-dimensional fictions where curved forms are not present. It has been suggested that for fractals where the aforementioned methods are not applicable, the box counting dimension method can be used to determine the fractal value [1, 28].

3.1 Box counting method

The box counting method is the most preferred mathematical method for calculating the fractal dimension. In this method, the calculation of the fractal size is based on the logic of calculating the number of boxes that contain data in a two-dimensional drawing. This is done by dividing the drawing into a grid system and determining the number of boxes that contain relevant information. In this method, as the details increase, the fractal dimension also increases, and the details such as window records, doors, floor moldings, façade decorations on the façades cause the fractal dimension to be high [29].

In the box counting method,

1. The surface for which the fractal size is to be calculated is covered with a grid system (**Figure 4**).

2. Grids within the grid system that contain data at any point are considered full and their number is determined.

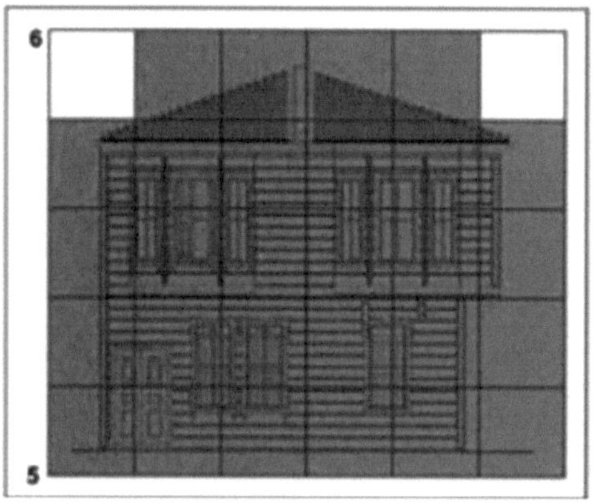

Figure 4.
Box counting method grid system [29].

3. To account for the observer's distance from the structure and the changing scale, the grid sizes in the grid system created for calculating the fractal size are adjusted. As the observer approaches the surface, the dimensions of the grids are halved, resulting in a new grid system. This process is repeated constantly, forming a cycle of grid adjustments.

4. During each cycle, the number of boxes containing data is calculated and noted.

5. $D = \frac{\log(x) - \log(y)}{\log(z) - \log(q)}$ The fractal value is calculated with the help of the formula (x: the number of boxes with data in it calculated in the next cycle, y: the number of boxes with data in the previous cycle, z: the number of boxes with data in the direction x in the next cycle, q: the number of boxes in the x direction in the previous cycle).

6. The fractal value is calculated using this formula for each cycle, repeating the process until the fractal value converges [1, 21, 29]. If the difference between the result decreases as the cycles continue, the process is terminated and the final value obtained is considered the fractal dimension.

The fractal value obtained by this method is always between 1 and 2, and it is stated that as you approach 1, the fiction is simple, plain and based on Euclidean geometry, and as you approach 2, the fiction is complex and irregular (**Figure 5**) [21, 30].

Within the scope of this study, in line with the purpose of the study, the Fractal theory, which provides the analysis of architecture with numerical data, was used for the analysis of the façades of the public buildings of the First National Architectural Period in Afyonkarahisar. In this direction, the box counting method, which is the most commonly used method for calculating the fractal dimension value of the façades, was preferred.

3.2 First National Architectural Period Public Buildings in Afyonkarahisar

Within the scope of the study, three public buildings belonging to the First National Architectural Period in Afyonkarahisar were selected. Fractal analysis of the façades of the Zafer Museum (1919–1920), the Rıza Çerçel Cultural Center (1930) and the Police Station Building (early 1900s) was performed using the box counting method. The architectural features of public buildings are as follows:

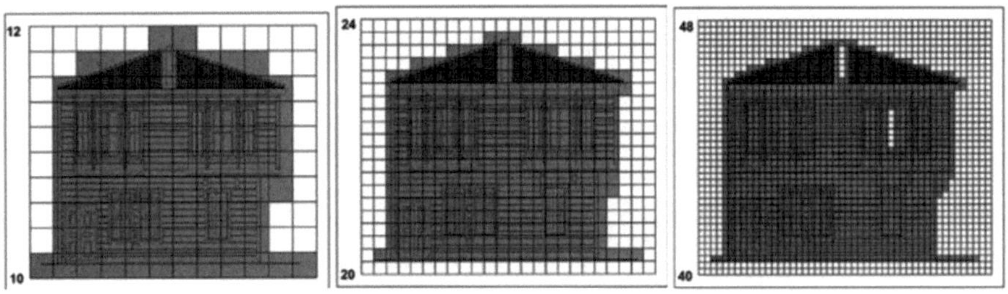

Figure 5.
Box counting method grid system grid size change [29].

3.3 Zafer museum

The building possesses the characteristic of being the first municipally-designed structure. While limited information is available regarding the architect of the building constructed in 1919–1920, it is stated that an Armenian architect was responsible for its construction [31, 32]. The building was registered as a cultural property to be protected by the High Council of Monuments on 11.07.1980 [33].

The building, located in the city square in the center of the city, was built as a two-storey masonry [34]. The main entrance is provided by the pointed arched structure facing the city square, and there are also entrances on the back and side façades of the building. On the side façade, there is a door accessed by a staircase at the corner of the first floor, and on the rear façade, there is an entrance obtained by transforming the window opening to the hall into a door. The presence of horizontal moldings on both sides of the main door, as well as between floors, is evident on all façades of the building [35]. The main and intermediate walls of the building are made of stone, the ceiling tiles and roof are made of wood. On the exterior of the building, the perimeter and corners of the entrance door located in the front are covered with andesite stone, the arched windows of the building, the windowsills and perimeter of the arched windows and the keystone on the top are covered with finely carved stone. The entrance eaves of the building are high and decorative ornamented woods are nailed to the eaves. The roof of the building is covered with Turkish style tiles (**Table 1**) [33].

The building hosted the Afyonkarahisar Congress in 1920, was used as the Greek headquarters until 25 August 1922, and as the command center of the Turkish army after the enemy occupation of Afyonkarahisar on 27 August 1922. The main decisions of the Dumlupınar War were made in this building. The building has a historical significance in this respect. Between 1922 and 1940 the building was used as municipality building, between 1963 and 1985 it is used as police department and 26 August 1995 it started to be used as the Zafer Museum [33].

3.4 Rıza Çerçel cultural center

The building, which was designed as the Turkish Hearth building, started in 1930, to be used as the branch of the Community Centers in Afyon on 19 February 1932. The projects of the building were prepared by Architect Arif Hikmet Koyunoğlu, the project designer of the Ankara Türk Ocakları service building [36].

The building, located on Ordu Boulevard in the center of the city, is a two-storey masonry building. The entrance of the building, where the main entrance is on the façade facing Ordu Boulevard, is provided by a nine-step staircase with a marble railing. The entrance hall is reached when you enter through the two-winged wooden entrance door. Located on the front, the room has a balcony, the front of which is covered with glass and iron joinery.

The roof of the building is hipped roof and covered with Marseille tiles.
The wide eaves of the building are decorated with laths by creating geometric patterns [34].

It is stated that Atatürk visited this building during his visit to Afyon on 2 March 1931 and wanted the building to be converted into a Community Center, where upon the building used as a Turkish Hearth was converted into a Community Center [36, 37]. The building was transferred to the Ministry of Finance upon the closure of the Community Centers and was used as the Revenue Office until 1992 [34]. With the establishment of Afyon Kocatepe University, it started to be used as the Rectorate

Zafer Museum	
 Zafer Muesum entrance façade figure [35].	
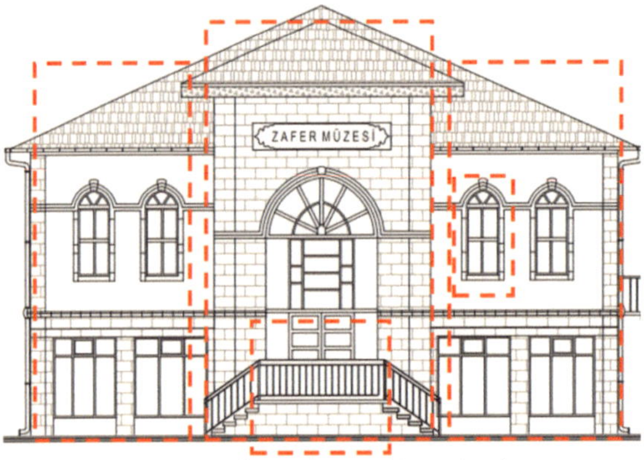 Zafer Museum entrance façade (Created by authors)	Façade analysis; • Gradations located horizontally and vertically, • The use of pointed arches on the windows, • The use of wide eaves, • Reading the floor traces from the façade with floor molding, • Raised entrance with steps, • The façade consists of three parts.

Table 1.
Zafer museum Façade analysis.

building, and today it has been used as a cultural center affiliated to Afyon Kocatepe University (**Table 2**) [38].

4. Yaylabağı police station building

During the early1900s, the Police Station Building was built with the purpose of ensuring security in Yaylabağı and the surrounding villages. It served this role effectively for many years. The building, which is open on four sides, was used as a police station, Greek headquarters and school in the historical process and was emptied and abandoned in the 1970s.

The two-storey building, which is masonry with a hipped roof, is made of cut stone and the windows are plastered on the brick. The building has a stone window and a flattened arch. On the ground floor of the building, there are two rooms and a staircase hall, one of which provides access to the structure, and on the first floor, there are two rooms and a staircase hall.

Rıza Çerçel Cultural Center

Rıza Çerçel Cultural Center entrance façade figure [39]

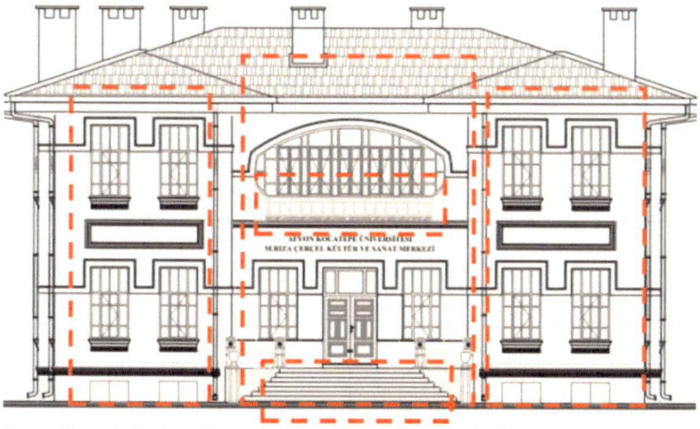

Rıza Çerçel Cultural Center entrance façade (Afyon Kocatepe University Department of Construction Archive)

Façade analysis;

- Gradations located horizontally and vertically,
- The use of eaves with wide and sub-surfaces,
- Reading the floor traces from the façade with floor molding,
- Balcony above the entrance,
- Raised entrance with steps,
- The façade consists of three parts

Table 2.
Rıza Çerçel cultural center Façade analysis.

In the examinations carried out in 2005, it was determined that there were collapses between the roofs and floors of the abandoned structure, the window and door joinery were broken, and the plaster was poured. Upon the application of Yaylabağı Municipality, the building was restored and re-functioned as a reading hall and library (**Table 3**) [37].

4.1 Fractal analysis of the Façades of the first national architectural period public buildings in Afyonkarahisar

Within the scope of the study, firstly, building façade survey drawings were prepared for the analysis of the façades of three public buildings belonging to the First National Architectural Period in Afyonkarahisar with the box counting method, and a 3×5 grid system was added on the façade drawing of each structure. In order to calculate the fractal value in the first cycle, the number of filled boxes was determined and the fractal value was calculated according to the formula. In the next cycles, the grid system was changed to 6×10, 12×20, 24×40, 48×80, respectively, and the number of filled-empty boxes for the façades is shown in the tables (**Tables 4–6**).

Yaylabağı Police Station

Yaylabağı Karakolu façade figure [37]

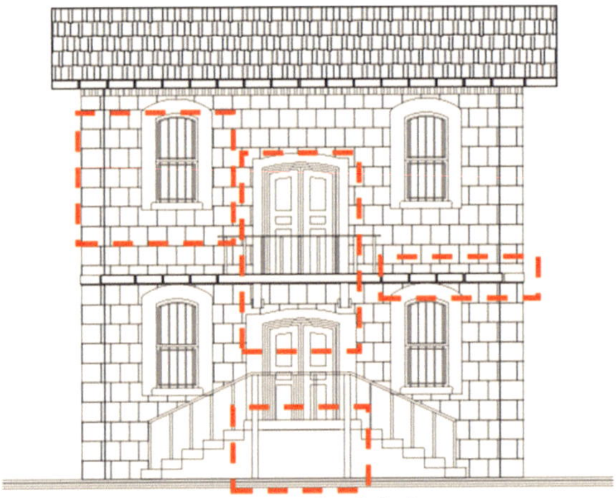

Façade analysis;

- Reading the floor traces from the façade with floor molding,
- Balcony above the entrance,
- Raised entrance with steps,
- Flattened arched windows

Yaylabağı Police Station entrance façade [37]

Table 3.
Yaylabağı Karakolu Cephe Analizi.

In line with the data in **Tables 4–6**, with the help of the formula, calculations $D = \frac{\log(x) - \log(y)}{\log(z) - \log(q)}$ were made for each building façade and the fractal dimension values resulting from the calculations were shown in **Table** 7.

As a result of the calculations, it was seen that the fractal dimension values of the three selected structures varied between 1 and 2. According to the values obtained as a result of 4 conversions; the building with the highest fractal dimension value is the Zafer Museum with a value of 1.93, while the building with the lowest fractal dimension value is the Police Station Building with a value of 1.16.

5. Conclusion

The objective of this chapter is to analyze the façade layout of three public buildings belonging to the First National Architectural Period in Afyonkarahisar by fractal method. In this direction, first of all, the architectural features of the First National

		Total number of boxes	Number of full boxes	Number of empty boxes
1. Conversion		15	13	2
2. Conversion	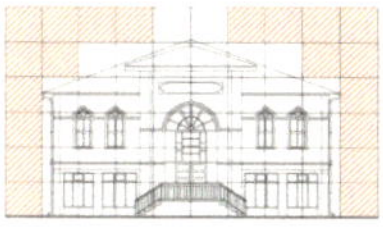	60	40	20
3. Conversion		240	148	92
4. Conversion	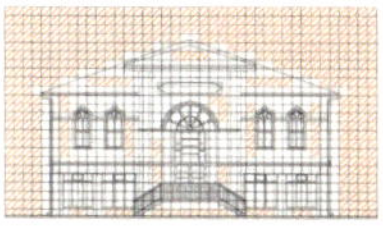	960	508	452
5. Conversion		3840	1873	1967

Table 4.
Zafer museum box count metrics.

		Total number of boxes	Number of full boxes	Number of empty boxes
1. Conversion		15	15	0
2. Conversion		60	60	0
3. Conversion	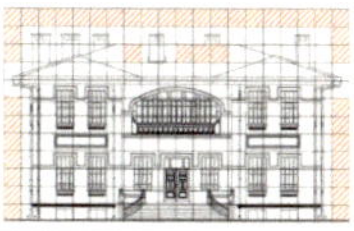	240	194	46
4. Conversion	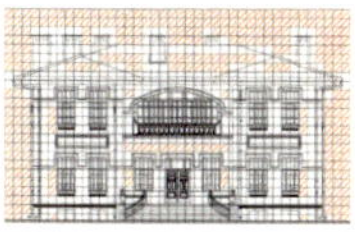	960	671	289
5. Conversion		3840	2073	1767

Table 5.
Rıza Çerçel cultural Center box count metrics.

		Total number of boxes	Number of full boxes	Number of empty boxes
1. Conversion		15	9	6
2. Conversion	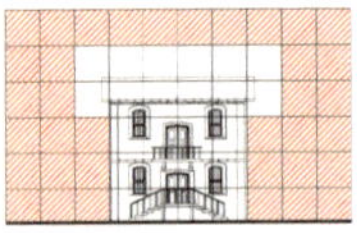	60	24	36
3. Conversion	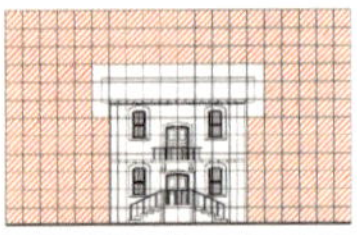	240	78	162
4. Conversion	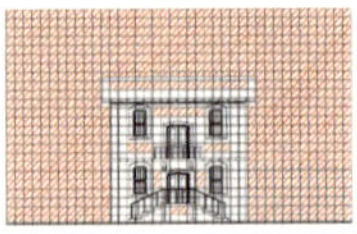	960	228	732
5. Conversion	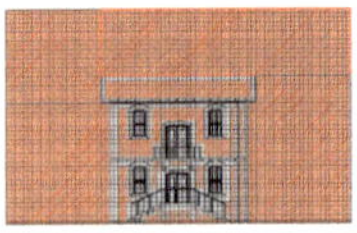	3840	624	3216

Table 6.
Police Station building box counting measurements.

	Zafer Museum	Rıza Çerçel Cultural Center	Police Station Building
1. Conversion	1.58	1.93	1.38
2. Conversion	1.90	1.71	1.70
3. Conversion	1.86	1.73	1.53
4. Conversion	1.93	1.72	1.16

Table 7.
Fractal dimension values of building façades.

Architectural Period were revealed in the study and the façade layouts of the three selected historical buildings were analyzed in the context of these features. Then, in order to obtain numerical data for the systematic analysis of the façade layouts, fractal size calculations of the façades were made by box counting method.

It has been seen that the structure with the highest fractal size is the Zafer Museum. During the analysis, it was determined that the entrance façade of the Zafer Museum exhibited distinct characteristics associated with the First National Architectural Period. These features included the presence of horizontal and vertical staggers, the use of pointed arches to emphasize window gaps, the inclusion of floor moldings that allowed the reading of floor traces from the façade, an elevated entrance accessible via stairs, a façade divided into three parts, and the utilization of variety of materials. These characteristics contributed to the overall composition and reading of the façade. In this structure, the proximity of the fractal size value to 2 has been interpreted as an indication of the intricate complexity of the façade layout and the presence of numerous intricate details.

The building with an average fractal size is the Rıza Çerçel Cultural Center. This situation belongs to the First National Architectural Period on the façade of the Rıza Çerçel Cultural Center. It was determined that features such as the presence of horizontal and vertical cascades, the use of eaves with wide and sub-surface decorations, an entrance reached by a raised staircase, the façade consisting of three parts and the use of uniform materials were related to the reading from the façade. In this structure, the interpretation of the fractal size value ranging between 1.5 and 2 suggests that the façade layout is relatively simple in its overall composition. However, it also indicates the utilization of intricate details within the design.

It was seen that the structure with the lowest fractal size was the Police Station Building. It was determined that this situation was related to the fact that the features of the First National Architectural Period in the Police Station Building such as the appearance of floor traces only with floor moldings, an entrance reached by a raised staircase, the balcony above the entrance, the flattened arched window spaces and the use of uniform materials were read from the façade. In the façade layout of this historical building, the features of the First National Architectural Period are less visible than the other buildings selected within the scope of the study. It was seen that this situation was also reflected in the numerical data obtained and the fractal dimension value of the façade of the building was lower than the other structures within the scope of the study. The interpretation of the fractal dimension value being close to 1 in this structure suggests that the façade has a simpler and more streamlined design compared to other structures. The value approaching 1 indicates a reduced level of complexity and intricacy in the façade layout.

As a result, it is seen that the features of the First National Architectural Period in the façade fictions of the buildings selected within the scope of the study cause mobility and complexity in the building facades and this is reflected in the numerical expressions of the façade fictions. It is concluded that the façade characteristics of the structures selected within the scope of the study coincide with the numerical data obtained in the study.

Author details

Gamze Çoban and Şerife Ebru Okuyucu*
Department of Interior Architecture and Environmental Design, Afyon Kocatepe University, Afyonkarahisar, Turkey

*Address all correspondence to: ebruokuyucu@hotmail.com

References

[1] Bovill C. Fractal Geometry in Architecture and Desing. Boston: Birkhauser; 1996

[2] Sözen M. Cumhuriyet Dönemi Türk mimarlığı. Ankara: Türkiye İş Bankası Kültür Yayınları; 1996

[3] Yavuz YI. Ulusal Mimarlık Dönemi Ve Mimar a. Kemaleddin Bey: Ankara; 1981. p. 21

[4] Aslanoğlu İ. Erken Cumhuriyet Dönemi Mimarlığı. Bilge Kültür Sanat: Türkiye-İstanbul; 2010. pp. 17-271

[5] Ertuğrul Z. Birinci Ulusal Mimarlık Dönemi Mimarlarından Muzaffer Bey: Eserleri Ve Sanat Anlayışı. [Yayınlanmamış Doktora Tezi]. Eskişehir: Sosyal Bilimler Enstitüsü, Anadolu Üniversitesi; 2007. p. 13

[6] Hasol D. Ansiklopedik Mimarlık Sözlüğü. Yapı Endüstri Merkezi: İstanbul; 1998. p. 465

[7] Sözen M. Birinci Ulusal Mimarlık Dönemi ve Mimarları. İstanbul: Mimaride Türk Milli Üslubu Semineri;, Türkiye İş Bankası Yayınları; 1984

[8] Yavuz Y. Cumhuriyet Dönemi Ankarası'nda Mimari Biçim Endişesi. Mimarlık. 1973;**11-12**:26-44

[9] Sıkıçakar A. Birinci Ulusal Mimarlık Dönemi Giriş Cepheleri Analizi, [Yüksek Lisans Tezi]. İstanbul: İstanbul Teknik Üniversitesi; 1991

[10] Alsaç Ü. Türkiye'deki Mimarlık Düşüncesinin Cumhuriyet Dönemindeki Evrimi. [Doktora Tezi]. Trabzon: Fen Bilimleri Enstitüsü. Karadeniz Teknik Üniversitesi; 1976

[11] Bozdoğan S. Modernizm Ve Ulusun İnşası, Erken Cumhuriyet Türkiye'sinde Mimari Kültür. İstanbul: Metis Yayınları; 2001

[12] Ünsal B. Mimarlığımız 1923-1950. Mimarlık. 1973;**122**:20-25

[13] Yaldız E, Parlak Ö. Birinci ulusal mimarlık kamu yapıları. Social Sciences Studies Journal. 2018;**4**(24):4930-4947

[14] Ural S. Türkiye'nin Sosyal Ekonomisi ve Mimarlık (1923-1960). Mimarlık. 1994;**123-124**:5-53

[15] Barnsley MF. Fractals Everywhere. Boston: Academic Press Professional; 1993

[16] Erzan A. Doğadaki Fraktallar. Bilim ve Teknik Dergisi. 1998;**365**(23):34-39

[17] Mandelbrot BB. The Fractal Geometry of Nature. New York: W. H. Freeman and Company; 1983

[18] Fraktal DB, Tasarım GVÜSM. Yüksek Lisans Tezi. İstanbul: İstanbul Teknik Üniversitesi; 2009

[19] Kaos GJ. Tübitak yayınları. Ankara: 2005

[20] Falconer K. Fractal Geometry. Mathemathical Foundationa and Aplications. New York: John Wiley& Sons; 2003

[21] Kanatlar Z. Fraktal Boyuta Dayalı Mimari Bir Analiz: Sedad Hakkı Eldem ve Konut Mimarisi. Uludağ Üniversitesi. Bursa: Yüksek Lisans Tezi; 2012

[22] Peitgen HO, Jürgens H, Saupe D. Chaos and Fractals, 1992. New York: Springer-Verlag; 1992

[23] İbrahim M, Krawczyk R. Generating Fractals Based on Spatial Organizations.

Chicago: Illinois Institute of Technology College of Architecture; 2000

[24] Alik B. Mimarlıkta Tasarlama Yöntemleri Ve Fraktal Tasarımlar Üzerine Bir İnceleme, [Yüksek Lisans Tezi]. İstanbul: Kocaeli Üniversitesi; 2015

[25] Natural Fractals [Internet]. 2023. Available from: https://iternal.us/what-is-a-fractal/ [Accessed: 20.04.2023]

[26] Natural Fractals [Internet]. 2023. Available from: https://fractalfoundation.org/OFC/OFC-1-1.html [Accessed: 20.04.2023]

[27] Natural Fractals [Internet]. 2023. Available from: https://users.math.yale.edu/public_html/People/frame/Fractals/Panorama/Nature/MountainsReal/Mountains9.gif [Accessed: 20.04.2023]

[28] Büyükemir, Karagöz E. Geleneksel Konutlarda Cephe Kurgusunun Fraktal Boyut Analizi Kullanılarak Karşılaştırılması: Isparta Örneği. [Yüksek Lisans Tezi]. Isparta: Süleyman Demirel Üniversitesi; 2019

[29] Gözübüyük G. Farklı Mimari Dillerde Fraktallere Dayalı Form Üretimi, [Yüksek Lisans Tezi]. İstanbul: İstanbul Teknik Üniversitesi; 2007

[30] Mimari E, Ö, Yaklaşım TFKDÜB. İstanbul: İstanbul Teknik Üniversitesi. [Doktora Tezi]; 2003

[31] Yüksel İ. Afyonkarahisar'da Canlanan Tarih. Afyonkarahisar: Yerel Tarih Araştırmaları Merkezi Yayınları; 2005

[32] Eski ÜM, Binası B, Müzesi Z. Beldemiz. Afyonkarahisar Belediyesi Yayınları. 1995;**40**:11-12

[33] Abi M.Ü. Detaylarda Afyonkarahisar. Azim Matbaacılık. Ankara. s.415-s.419; 2020

[34] Uyan M. Anadolu'nun Kilidi Afyon. Afyon Valiliği. s.357; 2004

[35] Afyonkarahisar İ, K, Müzesi Z. Nevşehir Üniversitesi Sosyal Bilimler Enstitüsü Dergisi. 2011;**1**:s.245-s.267

[36] Ilgar Y. Türk Ocakları Afyonkarahisar Şubesi. Taşpınar Dergisi. 2001;**3**:s.23-27

[37] Abi MÜ. Geçmişin İzinde Afyonkarahisar/Kaybolan Yapılar. Azim Matbaacılık. Ankara. 2020;**s.115**:s.115-116

[38] Özpınar H. Bir Zamanlar Afyonkarahisar. Afyonkarahisar: Afyonkarahisar Belediyesi Yayınları; 2014

[39] Rıza Çerçel Cultural Center. 2023. [Internet] Available from: https://www.afyonhaber.com/images/haberler/2021/07/riza-cercel-kultur-merkezi-yenilendi.jpg [Accessed: 11.03.2023]

Chapter 6

Fractal Characterization of Microstructure of Materials and Correlation with Their Properties on the Basis of Digital Materials Science Concept

Maxim Sychov, Andrey Chekuryaev and Sergey Mjakin

Abstract

The concept of Digital Materials Science supposes that materials are designed, fabricated, tested, studied, characterized, and optimized on the basis of digital technologies, including the analysis of fractal parameters (fractal dimension, lacunarity, scale invariance, Voronoi entropy, etc.) of materials' microstructure. Many classes of materials may be considered as composites: polymer composites with inorganic fillers, alloys containing nonmetallic inclusions (oxides, carbides, nitrides, intermetallic ones, etc.), ceramic materials with pores and sintering additives, etc. The analysis of composition-technology-structure-properties relationships for such non-ordered composite materials requires the development of numerical tools for the characterization of their structure, including the interposition of phases. This chapter presents several examples of the implementation of this concept, including the study of filler distributions in dielectric composites, interposition of phases in special ceramic materials, distribution of nonmetallic inclusions in additively manufactured stainless steel, and structural features of tungsten oxide-based electrochromic materials. Based on the analysis of such characteristics as lacunarity and surface functionality, interrelations are established between technical properties of the studied materials and their structure providing approaches to the prediction and optimization of their target performances.

Keywords: steel, composites, ceramics, lacunarity, scale invariance, structure, interposition, additive manufacturing

1. Introduction

"Industry 4.0" (4th industrial revolution) is a concept of design, fabrication, testing, characterization, and utilization of products on the basis of digital technologies and Artificial Intelligence (AI). Originated from mechanical engineering and machine building industry, this concept involves the implementation of advanced techniques for manufacturing products via processing of materials possessing a set of required properties. In turn, target properties of materials are determined by their composition, structure, preparation, and pretreatment conditions. To take into account these numerous factors, the Industry 4.0 paradigm should be further supported and supplemented by the concept of Digital Materials Science, including the development of automated methods for the design and synthesis of materials, as well as computational approaches to the design of novel generations of advanced materials and digital analysis of the composition-structure-properties relationships [1].

Digital Materials Science includes but is not limited to the computational materials science [2]. The development of materials with adjustable target properties requires the proper design, synthesis of the required precursors, fabrication of the material in the desired form, as well as its study and characterization. Particularly, this concept includes the following aspects:

1. Design of materials at the atomic level of the structure, involving the crystal structure [3, 4].

2. Modeling the processes of the synthesis of materials and formation of a given composition and microstructure. For example, a periodic microstructure in inorganic materials can be tailored via adjustable reaction-diffusion transformations [5].

3. Modeling the macrostructure and properties of materials including biomorphic design [6, 7].

4. Digital methods for the description of the composition and structure of material [8].

5. Digital methods for the analysis of data on the composition-structure-properties relationships using artificial intellect [9–12].

6. Digital methods for the design of equipment for the manufacturing of substances (materials) with the desired properties [13–15].

7. Digital technologies for the fabrication of substances and materials, including the application of advanced robotic techniques [16, 17].

8. Study of materials using modern digital approaches, including databases, 3D scanning, computed tomography, etc. [18–20].

9. Digital quality control and management including robust design and manufacturing [21–23].

The Industry 4.0 and Digital Materials Science concepts are implemented at the Department of Theory of Materials Science and Department of Chemistry, Physics and Biology of the Nanoscale State (based on the Institute of Silicate Chemistry of the Russian Academy of Sciences (ISC RAS), head of the Department—Academician of the Russian Academy of Sciences Vladimir Ya. Shevchenko) of the Saint-Petersburg State Institute of Technology in a number of educational and research programs, including a new bachelor degree program "Design, synthesis and application of nanomaterials", master degree programs "Nanomaterials for Industry 4.0" and "Materials and technologies of additive manufacturing", and double-degree program for PhD students with the University of Shizuoka (Japan).

Presently, numerous methods are developed for the design, fabrication, and testing of materials. However, positions (4) and (5) in the above list are difficult to fulfill because of the lack of methods to describe the structure of non-ordered materials and thus to derive composition-structure-properties relationships. The properties of substances are determined not only by their chemical composition but also by their structure, i.e., by the mutual spatial arrangement of structural elements (atoms, ions, functional groups, etc.). For the crystal structure, such nanoscale arrangement is well described by the crystal lattice parameters with a strictly ordered translation of the unit cell to the entire volume of the material. However, on a larger scale level relating to the microstructure of composite systems such as the distribution of filler particles in various composites, phases (intermetallic, carbide, etc.) in alloys, and ceramic particles in cermets, the structural order at the microlevel is not so definite, since such parameters as the number of the nearest structural elements (coordination number) and the distance between particles or fragments of a certain phase are essentially statistical. Therefore, the order and regularity of the arrangement of structural elements in such systems can be established on a relatively large scale using statistical approaches.

Certain parameters characterizing the microstructure are required because the properties of materials largely depend on the features of the structural elements arrangement, including the uniformity of their distribution, mutual positioning, and distances between them. Hence, it is necessary to determine the corresponding quantitative characteristics and their relationship with the composition of the material, conditions for its formation, and target properties. A possible approach to this problem is based on the statistical analysis of the distribution of structural elements of a material between its individual volume fragments with the determination of fractal parameters characterizing the structural order, spatial uniformity of certain structural elements, and degree of self-similarity (scale invariance) of the system with an increase in the scale level.

In the recent decades, this approach involving the determination of such parameters as fractal dimension and lacunarity has been applied to the characterization of various materials. Particularly, in [24–26], these parameters were determined for fractured surfaces in steels and aluminum alloys depending on rupture modes with different crack propagation rates and heat treatment conditions. The fractal dimension was found to almost linearly correlate with the plain strain fracture toughness of steel and regarded as a measure of this parameter [24]. It was generally established that the fractal dimension negatively correlates with the dynamic tear energy and the critical crack extension force would rise faster than Hall-Petch's $1/\sqrt{d}$ law with the decrease of grain size due to the increase of the true area of the irregular fracture surface. In [26], the relationships among the fracture regimes and surface roughness index were analyzed in terms of fractal parameters.

More recently, fractal dimension values were determined and analyzed for the formation of dislocations and their organization into grains or subgrains in correlation with the mechanical behavior of metals and alloys, particularly plasticity characteristics [27].

In [28], the analysis of fractal parameters was shown to provide an effective approach to the comparative characterization of porous structure in stainless steels. Particularly, the fractal dimension was found to correlate with the porous structure of such steels growing with the increase of porosity.

In [29], fractal dimension and lacunarity were determined as parameters characterizing the porosity, pore size distribution, and other textural characteristics of carbon samples by the analysis of scanning electron microscopy (SEM) images and mercury porosimetry data. The obtained results provided an effective approach to the characterization of structural uniformity or heterogeneity of porous materials.

The considered parameters can be also applied to the study of surface characteristics of materials. In [30], lacunarity parameters derived from SEM images of femtosecond laser-irradiated metals and characterizing the irregularity of mutual arrangement of certain areas were shown to successfully quantify the spatial texture of the studied images, thus providing a convenient means of reporting changes in surface topography with respect to changes in processing parameters. Furthermore, since lacunarity plots are highly sensitive to the different length scales present within a hierarchical structure, the lacunarity analysis can be considered as a powerful tool for quantitative characterization of multi-scale hierarchical surface topographies.

A similar approach was also applied in [31] to comparatively analyze the surface smoothness and roughness as a result of metal coating deposition onto brass plates. The fractal dimension and lacunarity of the surface SEM image were established to be relevant parameters depicting the quality of coating surface at significance level of 0.05.

Hence, an essential goal is the development of a general concept as well as reliable and reproducible procedures applicable to the study of fractal parameters and surface characteristics for different kinds of materials in order to predict and adjust their target properties and performances. Particularly, an important objective in this field relates to quantitative characterization of various composites with respect to a precise analysis and modification of their microstructure, taking into account a number of factors including the uniformity of filler distribution in the matrix as well as the characteristics of the filler surface responsible for specific filler-binder and filler-filler interactions.

This chapter presents the results obtained by our research group in the development, modification, and implementation of the considered approaches to polymeric, metallic and ceramic materials, organic and inorganic films, etc. Since all these kinds of materials can be considered as composites with certain inclusions like filler particles, pores, and nonmetallic compounds, microstructure of such composites sufficiently influences their electrical, mechanical, optical, and other characteristics. A general approach to the study of the considered materials in this respect involves the characterization of their microstructure followed by the determination of such fractal characteristics as lacunarity, Voronoi entropy in correlation with target properties. In addition to these fractal characteristics, we also introduced the *scale invariance* as a new parameter reflecting a self-similarity of the material microstructure at various scale levels, determined by a comparative analysis of lacunarity or other fractal parameter measured at the studied object division into segments of different size.

The developed approach can be applied for various classes of materials in order to analyze all kinds of their nonuniformity, particularly with respect to the chemical composition according to the data of various mapping techniques, microstructure studied by microscopy (e.g., optical and electron of atomic force) or tomography, crystal size, and other characteristics.

2. Methodology

The determination of microstructural order characteristics of the studied materials is based on a modified box-counting method involving the analysis of SEM images (obtained using a TESCAN Vega 3 installation) of the studied materials using ImageJ software package and involves the following steps:

1. Isolation of certain structural elements (e.g., filler particles or phase inclusions) and determination of the coordinates of their centers of mass.

2. Partition of the analyzed fragments of SEM images into square cells with certain side size *x* and counting the number of centers of mass in each of these cells. This operation was performed in the Excel 2016 spreadsheet editor using the "AND" function, which checks the correspondence of the coordinates of the centers of mass to the coordinates of the segments.

3. Statistical treatment of the obtained results, including the calculation of the average value, range, standard deviation, and lacunarity Λ (parameter characterizing the inhomogeneity of filling the space with the considered structural elements):

$$\Lambda = \left(\frac{\sigma}{\mu}\right)^2 \tag{1}$$

where σ and μ are the standard deviation and the average number of centers of mass of the studied objects in the segments under consideration, respectively.

A decrease in the value of this parameter corresponds to a decrease in the proportion of areas free of the filler particles and, accordingly, to an increase in the uniformity of filling the space [32, 33]. Eq. (1) is essentially the ratio of the deviation of the number of particles in a cell and the average number of particles in a cell, and the smaller the deviation, the smaller the differences in the number of particles in the cells, and the more uniformly they are located.

Lacunarity is a parameter characterizing the deviation of a geometric object, such as a fractal, from translational invariance, i.e., indicating how parts from different regions of a geometric object are similar to each other at a given scale. Geometric objects with low lacunarity are homogeneous and translation-invariant, since all dimensions of the gaps (distances between the studied particles, for example) are the same. In contrast, objects with a wide range of gap sizes are inhomogeneous, not translation invariant and have a high lacunarity.

In other words, lacunarity is a characteristic of the uniformity of the distribution of any objects (particles) in space, i.e., the lower its value, the more uniformly the particles are distributed [33].

The considered procedure can be repeated with the variation of the cell size x with a certain step in order to obtain the lacunarity spectrum, i.e., the dependence Λ *vs* x. Based on thus studied effect of the cell size (scale factor) on the lacunarity, the *scale invariance* of the system can be determined as a parameter characterizing the sensitivity of lacunarity to the change of scale. Generally, the lacunarity decreases with an increase in the cell size providing a greater "averaging" of the structural elements distribution [33].

The resulting dependences for 3D objects are almost ideally approximated by a power law according to the following general expression:

$$\Lambda = \mathrm{A} \bullet \mathrm{x}^{\mathrm{n}-3} \tag{2}$$

or in the logarithmic form:

$$\ln\Lambda = \ln A + (n - 3)\ln x \tag{3}$$

where A is the pre-power coefficient and (n – 3) is the exponent factor characterizing the intensity of the lacunarity change with the cell size. This value represents the rate of change in the heterogeneity of the system with the change in the scale level (cell size). Generally, the less significantly the properties of the system change with the scale changes, the more scale-invariant this system is and the higher its ability to maintain its properties when the scale of study changes. Obviously, the most homogeneous system features the least prominent changes with a change in the scale. Thus, the factor n can be considered as a characteristic of the scale invariance of the system indicating its homogeneity. Since n value is commonly below 3, its growth corresponds to less negative power factors (n – 3) and consequently less prominent lacunarity drop with the increase of cell size reflecting higher scale invariance, self-similarity, and structural order of a material.

Furthermore, the properties of materials are determined not only by the composition, structure, properties, and mutual arrangement of their components but also by their surface properties and interfacial interactions which also should be optimized to improve the target performances. According to the concept of the scientific school created by the corresponding member of the Russian Academy of Sciences V.B. Aleskovsky and subsequently developed by professors S.I. Koltsov, V.G. Korsakov, A. A. Malygin, V.M. Smirnov, and A.P. Nechiporenko et al., a solid substance or material can be considered as a combination of a carcass and surface active centers of various nature, particularly featuring different acid-base or donor-acceptor properties. This factor should be taken into account at the consideration of target properties of materials (including electrophysical and electro-optical performances), since surface centers reflect the imperfection of the structure of the surface of a solid and largely determine its interaction with the environment (particularly including interfacial interactions in composites). Therefore, certain relationships can be established between the nature of such centers, their acid-base (donor-acceptor) properties, concentration on the surface, and various properties of materials sensitive to the state of the surface (including the characteristics of filler distribution in composites affecting their target performances) in order to extend the possibilities for the prediction and adjustment of the properties of functional materials by methods of solid state chemistry. Then the considered problem can be addressed by the adjustment of the surface functional composition, particularly toward the enhancement of certain interfacial interactions, in order to prevent or promote aggregation processes, improve the compatibility of the components, adjust their mutual distribution, and eventually improve the target performances of materials.

The development of this research direction requires the study of quantitative characteristics of the solid surface using digital methods followed by the search and analysis of correlations between the obtained data and target properties using digital materials approaches.

The suggested approach to study the composition-structure-properties relationships generally involves:

- analysis of the considered microstructural order parameters (lacunarity and scale invariance) and/or surface characteristics for a series of similar materials differing in their composition (e.g., content of a certain component or additive);

- measurement of target performances for the studied series of materials;
- analysis of correlations between the microstructural/surface characteristics and target performances in order to reveal the effect of the structural order features upon the properties of materials and provide the background for their prediction and improvement.

The applications of the considered approach to different classes of materials are summarized below.

3. Application examples

3.1 Polymer-inorganic composites

In a series of our recent studies [8, 34, 35], the considered approach was applied to the characterization, modeling, and prediction of the properties of polymer-inorganic dielectric composites. Such materials featuring a combination of a high dielectric permittivity (*k*) and flexibility are applied in various electronic devices involving a flexible dielectric layer, such as flexible electroluminescent panels, displays, film capacitors, capacitive energy sources, and other devices and systems of the new generation based on "smart" materials [36]. Although the dielectric permittivity of the most of conventional polymers does not exceed 10, the presence of polar groups (cyanoethyl, carbonyl, and hydroxyl) in the polymer backbone can significantly increase this value without any deterioration in other performances. One of the most promising polymers in this respect is cyanoethyl ester of polyvinyl alcohol (CEPVA) involving all the considered functional groups [37, 38]:

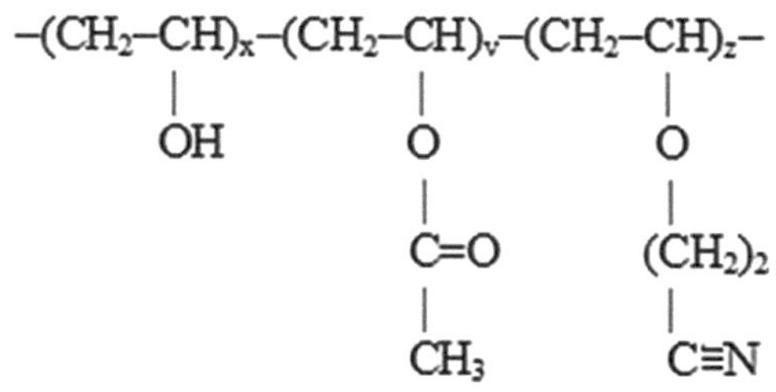

In order to provide the highest *k* values, composites based on such polymers should include various high-*k* disperse ferroelectric fillers, the most often used among which is barium titanate featuring one of the highest k values (more than 4000 for submicron particles).

In our previous studies [34, 35, 39], the filler surface modification with various nanolayers ("core-shell" structures) bearing specific Broensted acidic surface centers responsible for the interaction with the polymer binder via an acid-base mechanism was shown to provide a network of filler-polymer bonds, thus improving the components compatibility, preventing from the filler particles aggregation, and promoting the enhancement of dielectric performances. A similar effect can be achieved by the decoration of the filler surface with nanoparticles possessing the required active surface centers promoting the enhancement of acid-base interactions between the ferroelectric filler and polymer binder to provide the improvement of their compatibility and filler dispersion uniformity in the polymer.

In this study, barium titanate (Fuji Titanium, Japan; particle size about 0.5 μm, specific surface area 1.43 ± 0.07 m^2/g, dielectric permittivity more than 4000) was modified by the deposition of graphene nanoplates (RUSGRAPHEN, Ltd.) from 0.5% wt. colloid solutions in distilled water followed by evaporation, drying, and grinding to obtain $BaTiO_3$ samples containing graphene on the surface in the amounts from 0.2 to 6 mg per 1 g of barium titanate.

The surface functional composition of thus modified fillers was characterized by the adsorption of acid-base indicators selectively adsorbed on surface centers with certain pKa values corresponding to the transition between the acidic and basic forms of the indicator ($HInd \leftrightarrow H^+ + Ind^-$) with the change of its color (optical density at certain wavelengths). Spectrophotometric measurements of the change in optical density after the indicator adsorption as a result of the interaction with the studied material surface provide the determination of the contents of different adsorption centers on its surface [38–41].

The prepared modifies fillers were dispersed in 3 mL of CEPVA (PB paste, Shanghai Keyan Phosphor Technology Co, Ltd., Chine, k ≈ 25–30) in the amount corresponding to the filler content 20%vol. in the dry composite. Then the obtained mixtures were processed in an ultrasonic bath within 5 min to disperse the filler, followed by the deposition onto aluminum substrates via jets and drying at 70°C for 4 h, resulting in the formation of composite layers of about 50 μm average thickness.

Then electrodes comprising a silver-containing electrically conducting glue Contactol were deposited onto the composite layers, followed by measuring their capacitance and dielectric loss tangent (tgδ) using E7-20 immittance meter (MNIPI, Belarus) at frequency 1 kHz. The dielectric permittivity was calculated as:

$$k = \frac{C \cdot d}{\varepsilon_0 \cdot S}, \tag{4}$$

where C is the measured capacitance, and d and S are the composite layer thickness and Contactol electrode surface, respectively; $\varepsilon_0 = 8.85 \times 10^{-12}$ F/m is the dielectric constant. The errors of k and tgδ measurements according to the averaged data for 3–5 Contactol electrodes deposited onto the surface of each studied sample do not exceed 15% and 10%, respectively.

Examples of SEM images for the prepared composites with graphene contents 0, 0.6, 0.8, and 4.6 mg/g clearly illustrating the structural differences of the considered materials are presented in **Figure 1**. These data show that the addition of 0.6 mg/g (**Figure 1b**) graphene provides a certain improvement of the composite uniformity compared with a similar material without this nanocarbon additive (**Figure 1a**). On the contrary, the increase of deposited graphene amount to 0.8 mg/g $BaTiO_3$ (**Figure 1c**) significantly deteriorates the filler distribution uniformity and in the case of 4.6 mg/g (**Figure 1d**) graphene content a less prominent negative effect is observed. These results indicate a very high sensitivity of the composite structure to the additive content and suggest the necessity in a precise optimization of the composite composition to achieve the target performances.

The considered structural features clearly correlate with lacunarity calculated according to the above procedure and shown in **Figure 2** as a function of the SEM image division cell size. At relatively small (1–3 μm) cell sizes, the lowest lacunarity value is observed for the mostly structurally ordered composite with 0.6 mg/g. For the less uniform sample with graphene content 4.6 mg/g, a certain increase of lacunarity

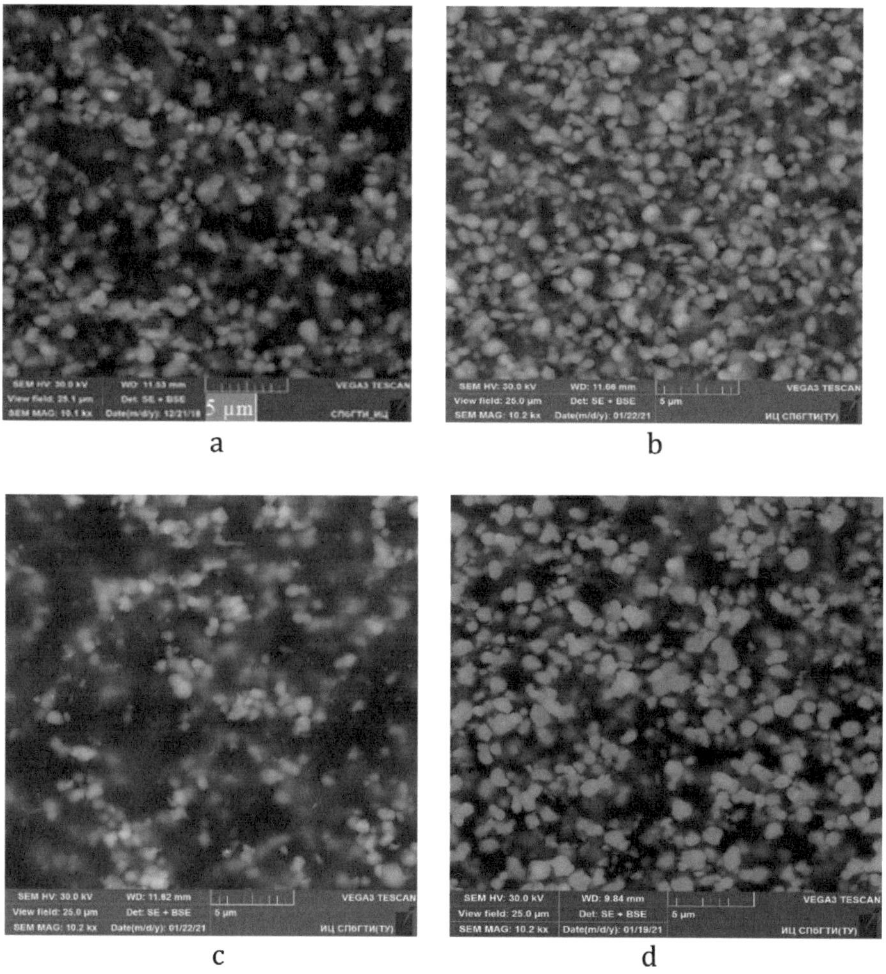

Figure 1.
SEM images of CEPVA-$BaTiO_3$-graphene composites with graphene additive contents 0 (a), 0.6 (b), 0.8 (c), and 4.6 mg/g $BaTiO_3$ (d).

is observed and the mostly disordered composite with graphene content 0.8 mg features the highest lacunarity. At higher cell sizes (4–6 μm), the applied method becomes less sensitive to the composite structure and lacunarity values drop to about 0.1.

The lacunarity *vs* cell size dependencies were analyzed for a whole series of composites with graphene content from 0.2 to 6 mg/g $BaTiO_3$ to determine the corresponding scale invariance parameters and compare the obtained results with the target performances of the synthesized materials. It was revealed that dielectric permittivity of the composites prominently grows with the increase (approaching zero) of the scale invariance parameter n (**Figure 3**).

The data in **Figure 3** suggest that the increase of permittivity in the range from less than 50 for graphene-free materials to about 125 for the composite with the optimal graphene content corresponds to the composite structure invariance to the scale factor, i.e., independence on the size of cells or fragments analyzed at the lacunarity determination.

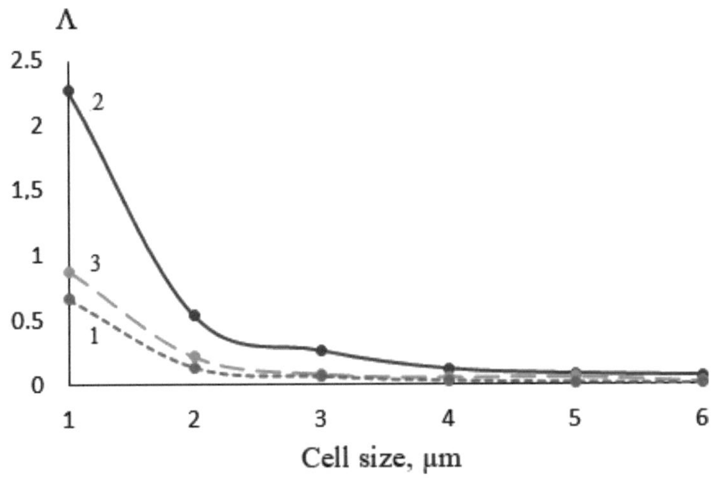

Figure 2.
Lacunarity of CEPVA-BaTiO$_3$-graphene composites with graphene additive contents 0.6 (1), 0.8 (2), and 4.6 mg/g BaTiO$_3$ (3) as a function SEM image division cell size.

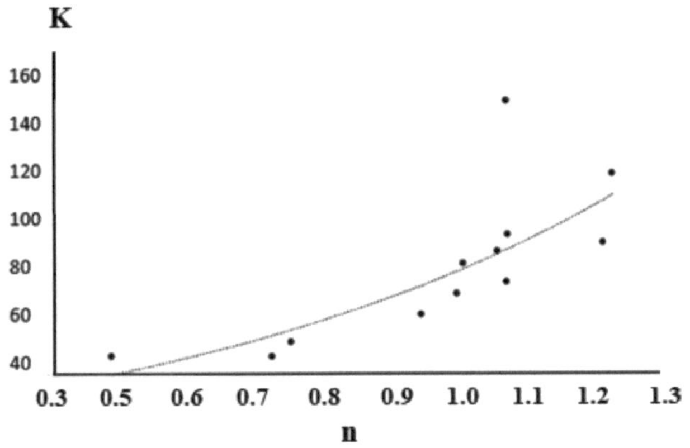

Figure 3.
Dielectric permittivity of CEPVA-BaTiO$_3$-graphene composites as a function of scale invariance parameter n.

The highest dielectric performances for CEPVA-BaTiO$_3$ composites (about 400) in couple with an almost 25% decrease in dielectric loss tangent (tgδ) was achieved in the case of barium titanate modification with fullerenol $C_{60}(OH)_{42}$ in the optimized amount of 3.2 mg/g BaTiO$_3$.

The analysis of corresponding SEM images (**Figure 4**) processed according to the considered approach indicated that the filler modification with optimal amount of fullerenol resulted in an almost double decrease in lacunarity from 0.15 to 0.08, suggesting an increased uniformity of the mutual arrangement of the filler particles without their aggregation [8].

This improvement of structural uniformity and permittivity of modified composites is determined by the presence of numerous hydroxyl functional groups in fullerenol capable of chemical interaction with OH-groups of CEPVA binder, thus preventing from the filler particles aggregation (**Figure 5**).

The involvement of surface functional groups in the observed enhancement of dielectric permittivity is confirmed by its prominent correlation with the content of

DOI: http://dx.doi.org/10.5772/intechopen.1002602

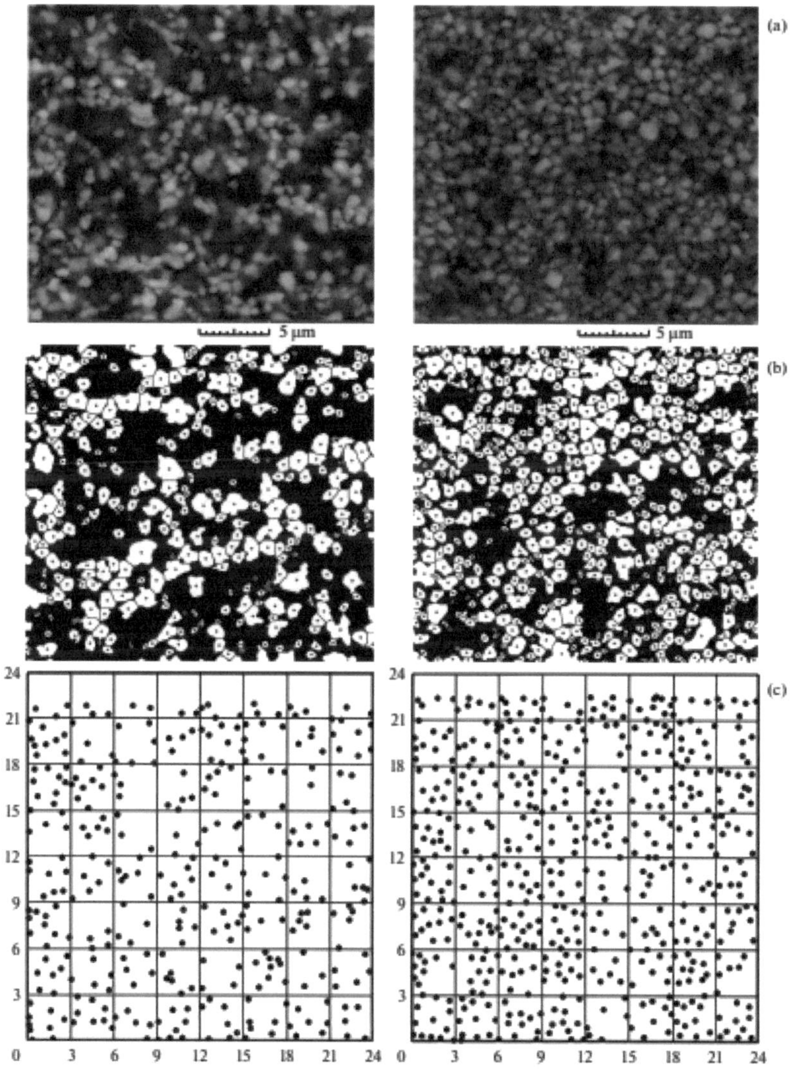

Figure 4.
SEM images (a), their binarized representation with marked centers of mass of particles (b), and the distributions of centers of mass of particles over 3 × 3 μm square cells for composites with the additive-free (left) and fullerenol (3.2 mg/g) modified $BaTiO_3$ filler (right).

Broensted acidic centers with pKa 1.3 supposedly corresponding to OH-groups in fullerenol (**Figure 6**). The optimal fullerenol amount provides the highest concentration of these centers on the fillers surface, while excessive additive content probably results in aggregation of fullerenol molecules via the condensation of OH-groups, thus preventing from their effective interaction with $BaTiO_3$ particles.

It should be particularly noted that the increase of fullerenol content above the established optimal level 3.2 mg/g results in a drastic decrease of permittivity probably as a result of undesirable interactions between the filler particles resulting in their aggregation.

Generally, the considered results demonstrate an efficient approach to precise optimization and prediction of the target performances of composites on the basis of quantitative relationships between the contents of their components, their

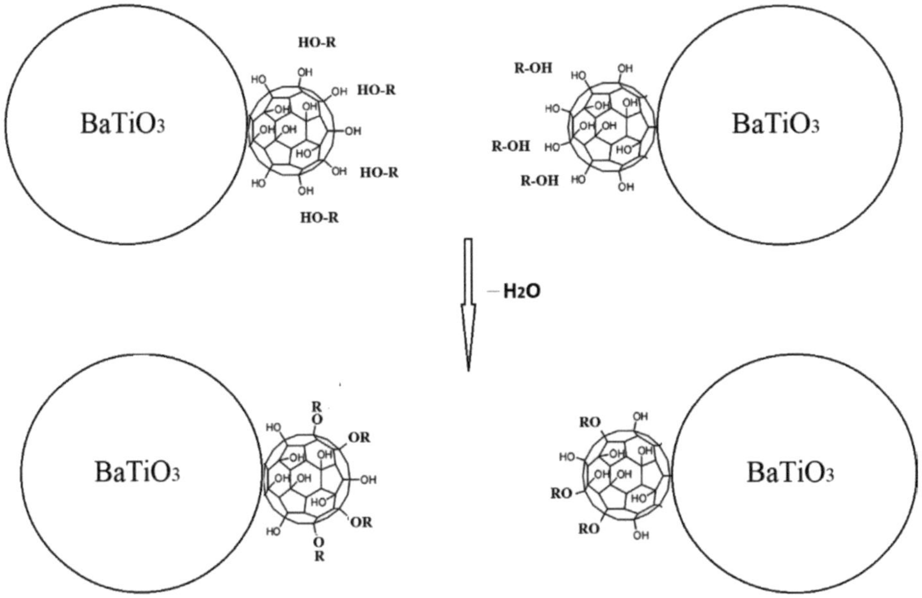

Figure 5.
Interaction of polymer binder R-OH with multiple OH-groups of fullerenol additive.

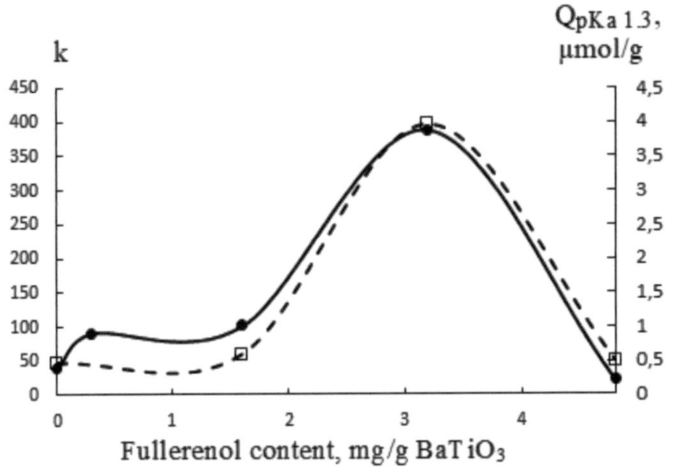

Figure 6.
Dielectric permittivity (——) of CEPVA-BaTiO$_3$-fullerenol composites and content of Broensted acidic centers with pKa 1.3 on the modified BaTiO$_3$ surface (——) as a function of fullerenol content.

microstructural features (characterizing their mutual arrangements) as well as their surface characteristics (particularly content of specific surface centers or functional groups) responsible for their interactions and compatibility.

3.2 Structure of composite ceramic materials

In [42], the developed methods were applied to a novel ceramic composite "Ideal" comprising diamonds distributed in a silicon carbide matrix and featuring outstanding

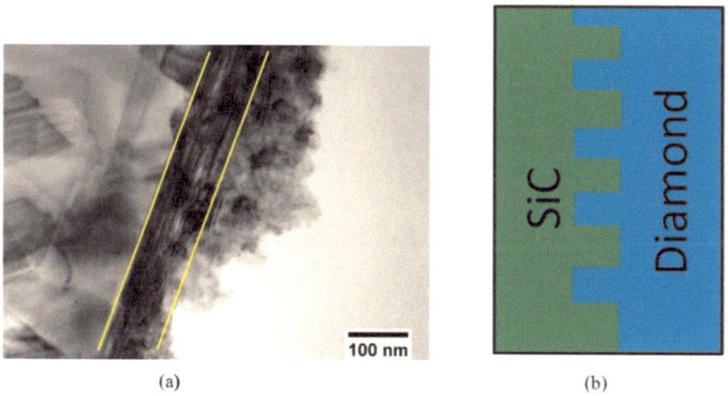

Figure 7.
Interpenetrating phase structure image (a) and schematic representation (b) for diamond-SiC nanocomposites (SiC is on the left, diamond—on the right side of the image, area of interpenetrating phases highlighted in yellow).

properties (e.g., modulus of elasticity 754 GPa significantly exceeding the available counterparts) due to both the composition and regular interconnected microstructure (triply periodic structures of silicon carbide on diamond particles formed during the reaction-diffusion synthesis involving filling the pore space between the diamond particles with silicon carbide, thus providing dense diamond-silicon carbide composite). The structure of this novel material is illustrated in **Figure 7**, and processing of its SEM image is shown in **Figure 8**.

In addition to the lacunarity analysis, this system was also studied by the determination of Voronoi entropy characterizing the probability *Pi* for the appearance of a system in the state *i*. Based on this concept, SEM images of the studied composites were used to plot Voronoi diagrams involving centers of mass of the filler particles as a finite set of points S.

Furthermore, since the studied materials consists of two fractions of particles greatly differing in size, it was decided to analyze both lacunarity and Voronoi entropy separately for the large (about 200–250 μm) or only small (below 28 μm) particle fractions in order to evaluate their contributions to the overall order and uniformity of the system.

The image illustrating the triangulation used for Voronoi entropy determination is shown in **Figure 9**.

Then the analyzed image was subjected to a partition into fragments involving a set of points located closer to one of the points of the overall set S than to any other point of the set [43]. After constructing the Voronoi diagram, we obtain data on the number of sides of the polygon built around each point; thus, we can subdivide the image into different classes of cells according to the number of sides (equal to the number of neighbors). The information entropy is determined as:

$$S(\varepsilon) = -\sum p_i \bullet \ln p_i, \tag{5}$$

where p_i is the probability that the considered Voronoi polygon has a certain number of sides. Generally, the Voronoi entropy indicates the amount of information required to determine the location of objects relative to each other and grows with the increase of the system disorder.

In this study, Voronoi diagrams were plotted using MATLAB software [44] involving the "voronoi" function. Also, using the "delaunayTriangulation" function, a

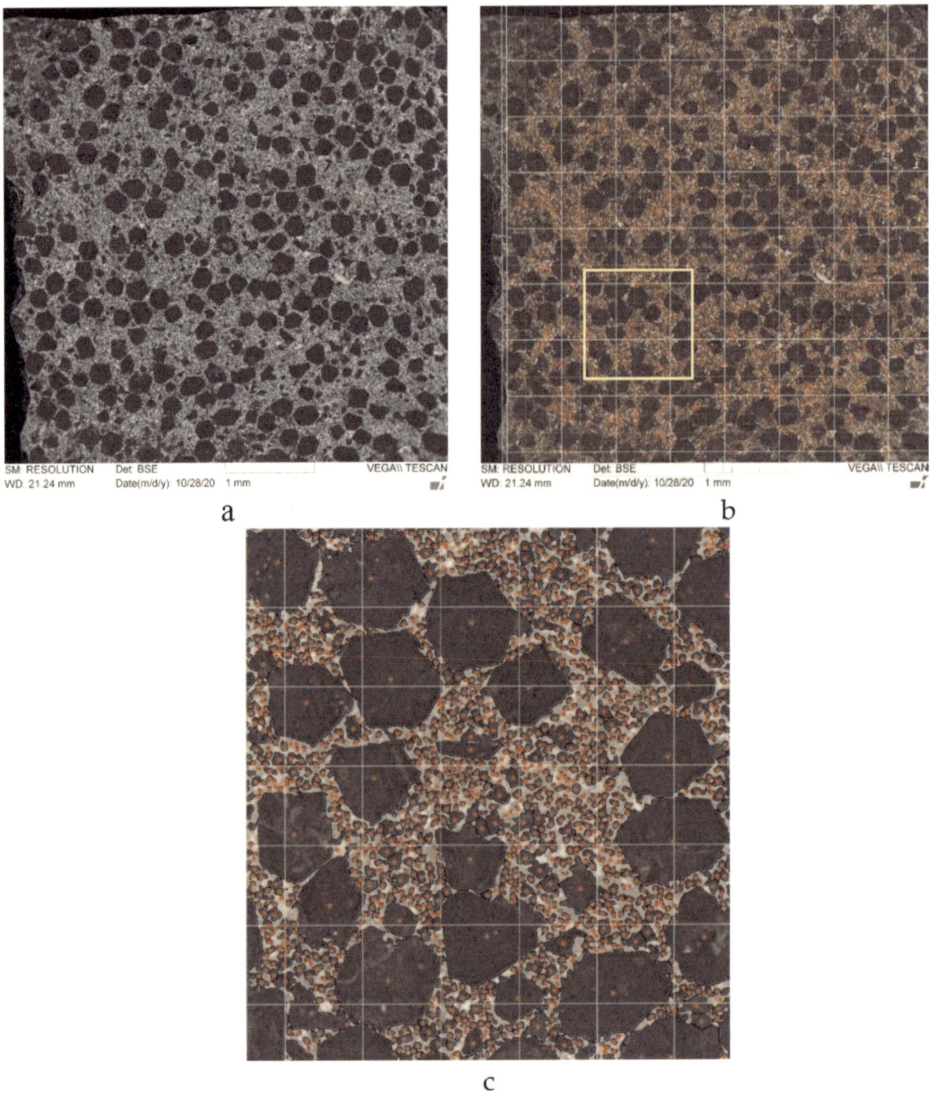

Figure 8.
4600 × 4600 μm-sized SEM image (140× magnification) of "ideal" composite (a), its division into 200 μm square cells including the centers of mass of the selected particles (b), and enlarged area highlighted in yellow square (c).

Delaunay triangulation was arranged for a given set of points on a plane, with all the circles described around any triangle contain inside no points except the triangle vertices. Based on this triangulation arrangement, Voronoi entropy was determined according to the procedure described in [45] MATLAB displaying the coordinates of the vertices of all Voronoi cells for each center of mass, indicating number of sides in the corresponding polygon is displayed as well as the coordinates of the start and end points of all triangulation lines. Furthermore, the applied software allows the determination of the number of nearest neighbors (coordination number) for each particle as the number of vertices of the sides in the Voronoi cell corresponding to this particle.

The obtained results summarized in **Table 1** indicate that the largest lacunarity is observed for the system of large particles (2.58), most likely due to a small number of

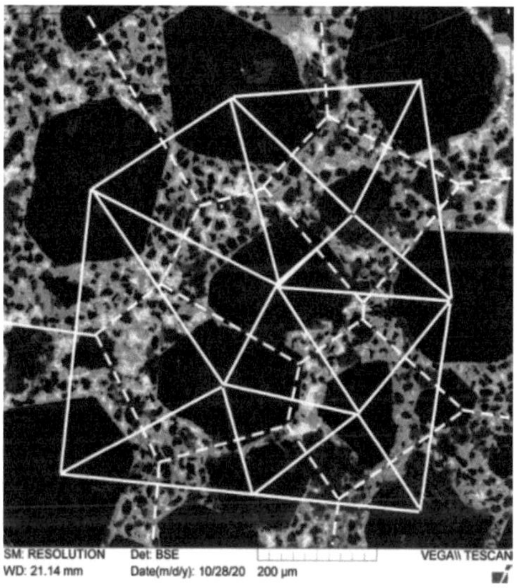

Figure 9.
Visualization of the Voronoi diagram (the sides of the Voronoi polygons are indicated as dashed lines) and triangulation (straight lines connecting particles) for the system of large particles.

Fraction	Lacunarity	Voronoi entropy
All particles	0.36	1.65
Only small (below 28 µm) particles	0.38	1.67
Only large (200–250 µm) particles	2.58	1.48

Table 1.
Lacunarity values calculated for different particle fractions.

such particles that cannot occupy all the cells, thus leaving many empty cells free of the particles, in turn causing a significant inhomogeneity. Probably, to study only large particles, it is necessary to consider images with a much larger coverage in order to obtain more reliable statistical information.

For the system involving only small particles, the lacunarity is much lower (0.38) indicating a more uniform distribution. Despite the presence of inhomogeneous regions in the place of eliminated large particles, the number of empty cells is significantly reduced, and the difference in particle filling density between different cells is less prominent.

The system involving both large and small particles features the lowest lacunarity (0.36) slightly below the value for the system of small particles that reflects filling of some empty cells with large particles.

The lowest Voronoi entropy is observed for the system of large particles, the highest value is calculated for small particles, and the system involving all the particles features a little lower entropy. The entropy minimum in the case of only large particles is determined by the decreased number of Voronoi polygon variants in terms of the number of sides in the absence of small particles. The increase in entropy in the case of a system including only small particles can be caused by an increase in the number of sides of the polygons associated with particles located next to the voids that were occupied by excluded large particles. The appearance of a point in the center of a

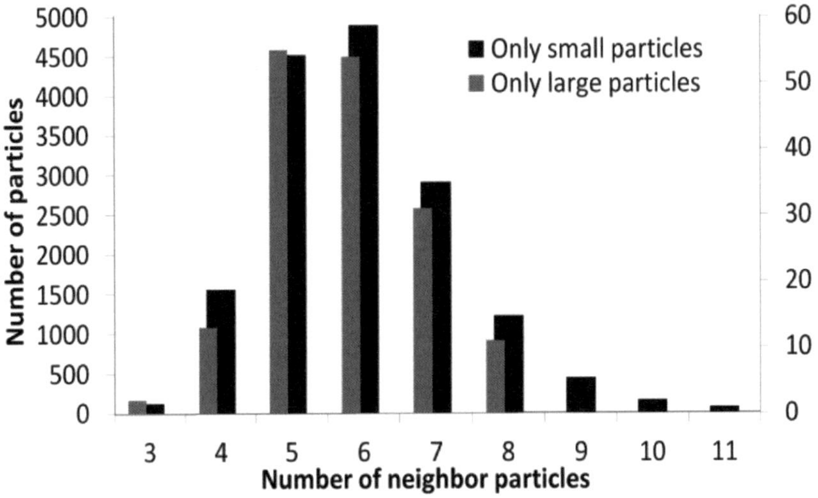

Figure 10.
Distribution of particles by the number of nearest neighbors for different fractions.

large void forms a Voronoi cell and the "extra" sides from the nearest small cells are "cut off."

Thus, the Voronoi entropy can be used for a comparative characterization of systems differing in the content of a filler (discrete phase distributed in a continuous phase) or structural order.

The number of nearest neighboring particles calculated for the systems of large and small particles is summarized in **Figure 10** as a histogram indicating the distribution of the number of particles with certain number of neighbors.

The maximum in this distribution for all particles and for the small fraction is 6 corresponding to the closest package. For the system of only large particles, the most probable number of neighbors is 5 and no values larger than 8 is observed.

Thus, the determination of Voronoi entropy provides an additional information such as the number of neighbors for each particle and distances between the neighboring particles that is promising for the analysis of the structure of various materials.

With respect to the target performances, the considered material features a clear correlation of the porosity and elasticity modulus with the scale invariance (**Figures 11** and **12**). These data suggest that the increase (approaching zero) of the scale invariance parameter n indicating the reduction of structural changes in the material with scale changes (i.e., increased scale invariance and structural order) corresponds to the reduction of porosity and enhancement of strength performances.

Hence, textural and mechanical characteristics of materials can be predicted on the basis of their microstructure analysis.

3.3 Tungsten oxide-based electrochromic materials and devices

In this case, microstructural order parameters were determined for tungsten oxide (WO_3) electrochromic layers deposited onto glass supports with fluorinated tin oxide (FTO) electrically conducting layers by magnetron sputtering using a novel high-throughput technique based on a periodic modulation of the deposition angle (PMMDA) [46, 47]. This method involves the rotation of the substrate around the axis perpendicular to the plasma flow direction to provide a variable angle of the sputtered

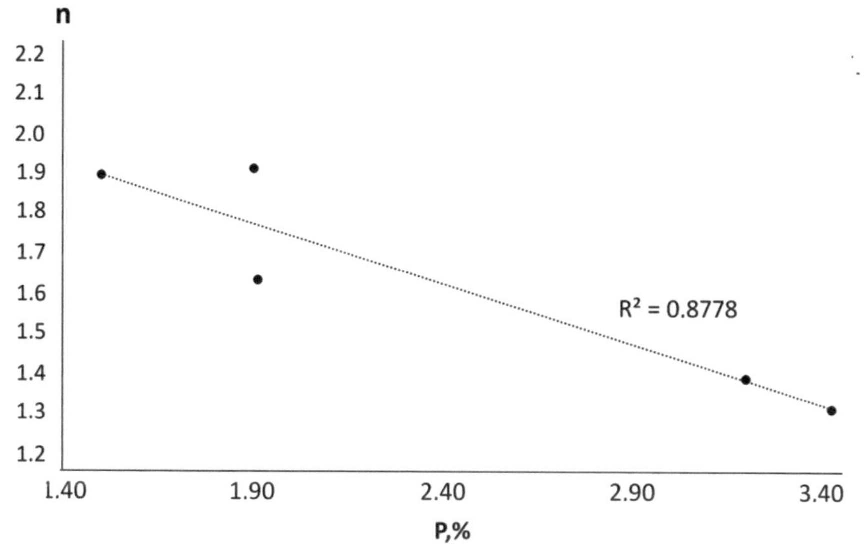

Figure 11.
Scale invariance factor n of "ideal" composites as a function of overall porosity.

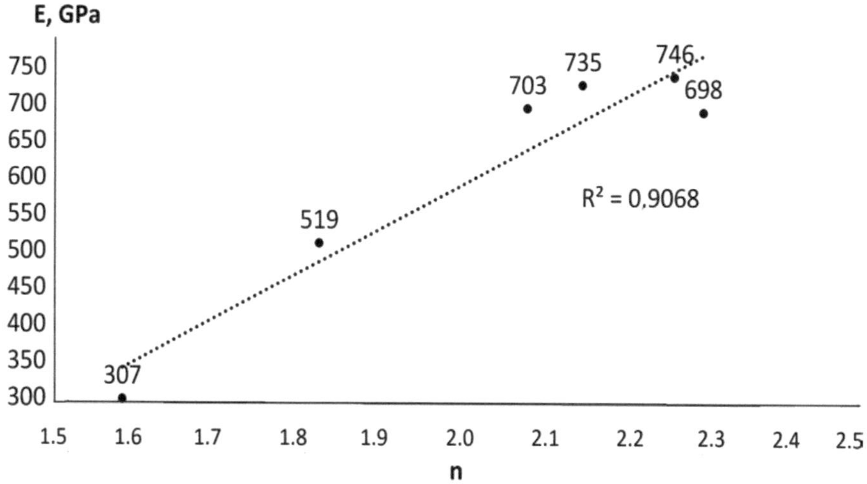

Figure 12.
Elasticity modulus of the "ideal" composite as a function of scale invariance factor n for medium and small fractions.

substance incidence and periodic exit of the substrate out of the flow. The key parameter responsible for the improvement of the target performances of electrochromic devices based on thus deposited layers (**Figure 13**) is the support rotation rate which determines an adjustable modification of the surface morphology and consequent increase of the contact area between the electrochromic layer and electrolyte (lithium perchlorate $LiClO_4$ solution) to improve the adjustable coloration and bleaching upon the application of electric current due to a reversible red-ox reaction:

$$WO_3 \text{ (colorless)} + xe^- + xLi^+ \leftrightarrow Li_xWO_3 \text{ (blue)} \tag{6}$$

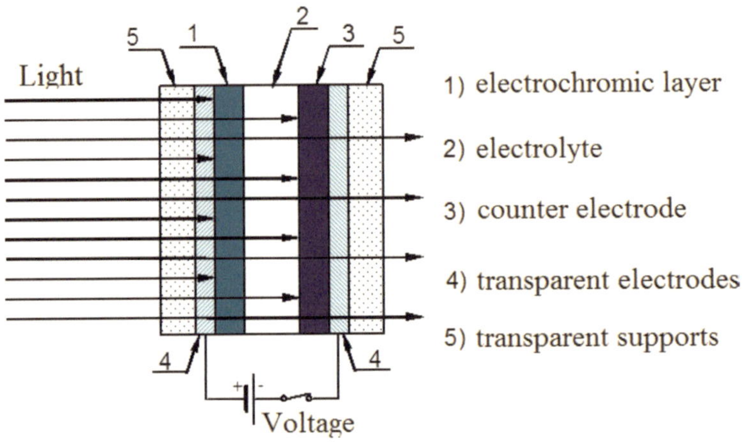

Figure 13.
Scheme of WO_3-based electrochromic device.

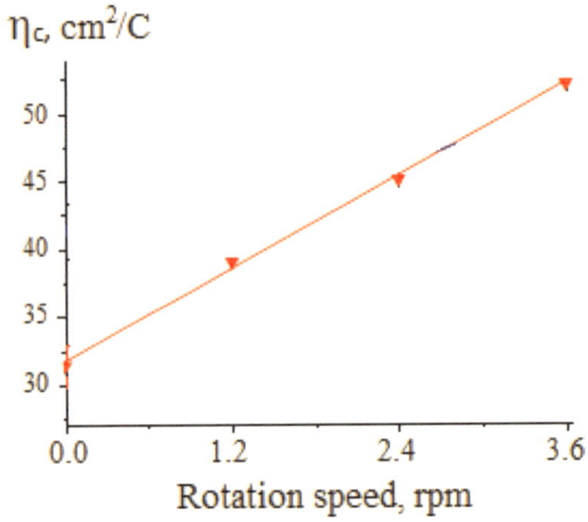

Figure 14.
Coloration efficiency of tungsten oxide-based electrochromic devices as a function of the substrate rotation speed at WO_3 layer magnetron deposition.

The increase of the support rotation speed up to 3.6 rpm resulted in a linear growth of the main target parameter—electrochromic coloration efficiency (**Figure 14**) determined as:

$$\eta_c = \frac{(D_{c2} - D_{b1}) * S}{\int I(\tau)d\tau} = \lg\left(\frac{T_{b1}}{T_{c2}}\right) * \frac{S}{\int I(\tau)d\tau} \tag{7}$$

where T_{b1}, T_{c2}, and T_{b2} (D_{b1}, D_{c2}, and D_b)—transmission (optical density) after bleaching at the 1st (previous) coloration-bleaching cycle, coloration at the 2nd (next) cycle, and bleaching at the 2nd (next) cycle, respectively, S is the sample surface, cm^2, and I is the current values measured every 1/3 s during the process time τ.

The surface morphology of tungsten oxide layers deposited in a stationary mode (without support rotation) and at different support rotation speeds was studied by atomic force microscopy (AFM) using a Ntegra Aura (NT-MDT, Russia) microscope with a scanning area of 5 × 5 μm. The obtained images (**Figure 15**) were processed by the box-counting method by division into 25 square 1 × 1 μm fragments and calculating the number of grains in each fragment followed by the calculation of lacunarity as described above.

The comparative analysis of the obtained results indicated that for the samples involving WO_3 layers deposited onto the rotating support, electrochromic efficiency almost linearly grows with the decrease in lacunarity (**Figure 16**).

3.4 Microstructure-properties relationship for stainless steels

The considered digital approach was applied to characterize the effect of heat treatment onto microstructural features of austenitic stainless steel X6CrNiTi18-10 samples produced via additive manufacturing (3D printing) technique by a selective laser melting method (SLM) using EOS apparatus [48].

Three types of samples were studied:

1. Initial steel without heat treatment as reference samples;

2. Samples subjected to heat treatment at 300°C within 60 min followed by cooling in air;

3. Samples subjected to heat treatment at 1050°C within 30 min followed by cooling in air.

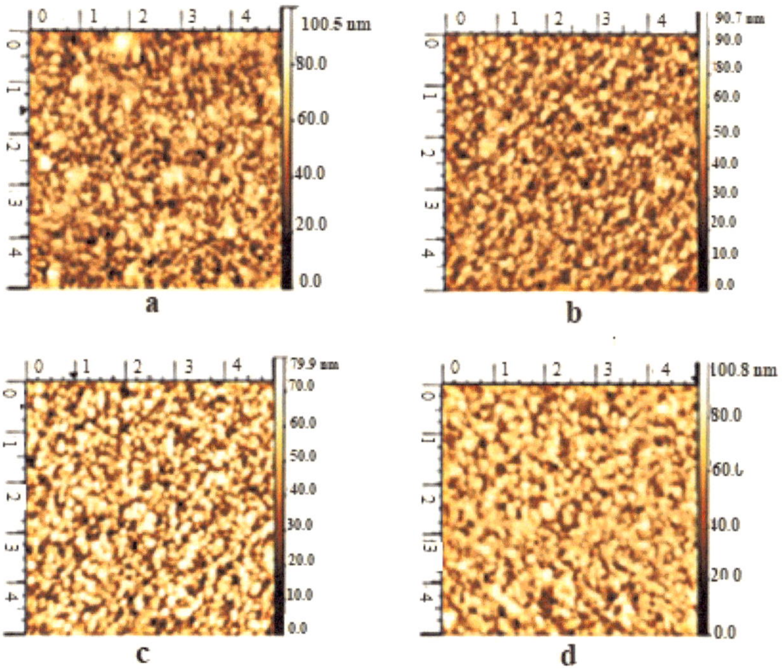

Figure 15.
AFM images of the samples deposited in a stationary mode (a) and onto the substrates rotating with the speeds 1.2 (b), 2,4 (c), and 3.6 rpm (d).

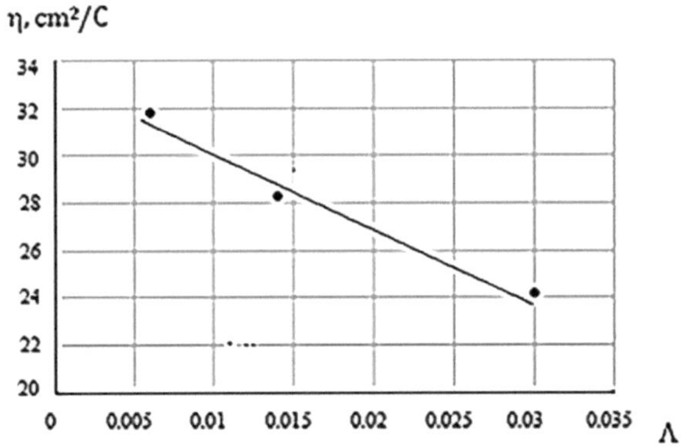

Figure 16.
Coloration efficiency of electrochromic devices involving WO_3 layers deposited with (•) and without (o) the support rotation as a function of lacunarity.

The applied heat treatment regimes are conventional for X6CrNiTi18-10 steel.

In order to investigate the influence of heat treatment on the steel microstructure, SEM images were obtained for cross sections of the studied 3D-printed samples.

The chemical composition of the obtained samples was studied using an atomic emission spectrometer DFS-50, X-ray fluorescence spectrometer LabCenter XRF-1800, as well as a carbon and sulfur analyzer CS-600 (LECO) in accordance with Russian Standards 12344-2003 and 12345-2001.

The quantity of nonmetallic inclusions was investigated according to Russian Standard 1778-70. The determination of these inclusions and analysis of their types was performed visually via microscopy.

The effect of heat treatment on mechanical performances of the studied steel samples was characterized by measuring their microhardness.

The elemental composition of the studied steel samples is shown in **Table 2**.

Microhardness of the obtained 3D steel samples is comparatively illustrated in **Table 3**.

Steel	Content of elements, % wt.							
	C	Cr	Ni	Ti	P	S	Si	Mn
Steel sample obtained by 3D printing	0.026	16.90	9.00	0.08	<0.01	0.015	0.31	0.58

Table 2.
Chemical composition of the original workpiece and steel sample after 3D printing.

Sample	Microhardness, $HV_{0.1}$, GPa
Initial 3D-printed steel without heat treatment	3.29 ± 0.16
Annealing at T = 300°C, 60 min	2.82 ± 0.21
Annealing at 1050°C, 30 min	2.57 ± 0.11

Table 3.
Microhardness of 3D-printed X6CrNiTi18-10 steel samples depending on heat treatment conditions.

Sample	Point oxides	String oxides	Silicates	Sulfides	Nitrides and carbides	Total
Initial 3D-printed steel without heat treatment	1.9	0.9	0	0	0	2.8
Annealing at T = 300°C, 60 min	1.7	1.1	0	0	0	2.8
Annealing at 1050°C, 30 min	2.3	0.6	0	0	0	2.9

Table 4.
Content of nonmetallic inclusions determined according to a standard procedure.

Since chemical composition of steel remains the same, the observed changes of mechanical properties obviously result from changes in the steel microstructure. The most significant influence on steel hardness has the content of nonmetallic inclusions (oxides, carbides, and nitrides) as their hardness is much larger than that of metal matrix. The amount of nonmetallic inclusions in the studied steel samples estimated in accordance with standard procedure is summarized in **Table 4**.

However, it is difficult to directly correlate the content of such inclusions with the steel hardness, since in the course of 3D printing a laser beam selectively melts steel particles one by one, thus forming melt bath with nonmetallic phases distributed along edges of the melted zone. After solidification, a 3D network of nonmetallic inclusions is formed, preventing from the movement of dislocations and mechanical deformations and thus providing increased mechanical properties. Therefore, the design of 3D network of nonmetallic inclusions is a promising approach to the enhancement of steel mechanical properties, and a detailed study of the achieved effects requires the application of more exquisite method to characterize the microstructure of 3D-printed steel.

SEM images of thin sections of samples annealed at different temperatures with distinguished centers of mass are shown in **Figure 17**. It was found that in 3D-printed samples nonmetallic inclusions are mainly concentrated in the space between crystallized particles of the remelted powder. The chemical composition of the inclusions suggests that they predominantly consist of titanium oxides.

The analysis of these data revealed that the increase in the annealing temperature leads to the increase in the lacunarity of nonmetallic inclusions (i.e., a growth of their distribution inhomogeneity), in couple with a decrease in their number and in the scale invariance factor n (**Figure 18**), reflecting a decrease in the statistical nature (stability) of the system.

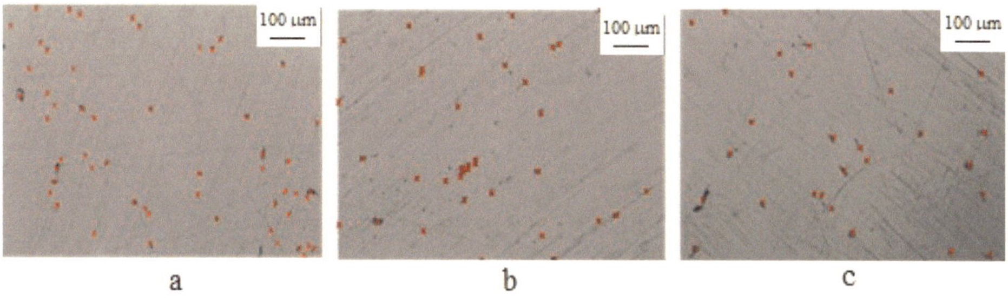

Figure 17.
SEM images of 3D-printed (SLM) samples from steel X6CrNiTi18-10 without heat treatment (a), annealed at 300°C (b) and 1050°C (c) with nonmetallic inclusions marked in red.

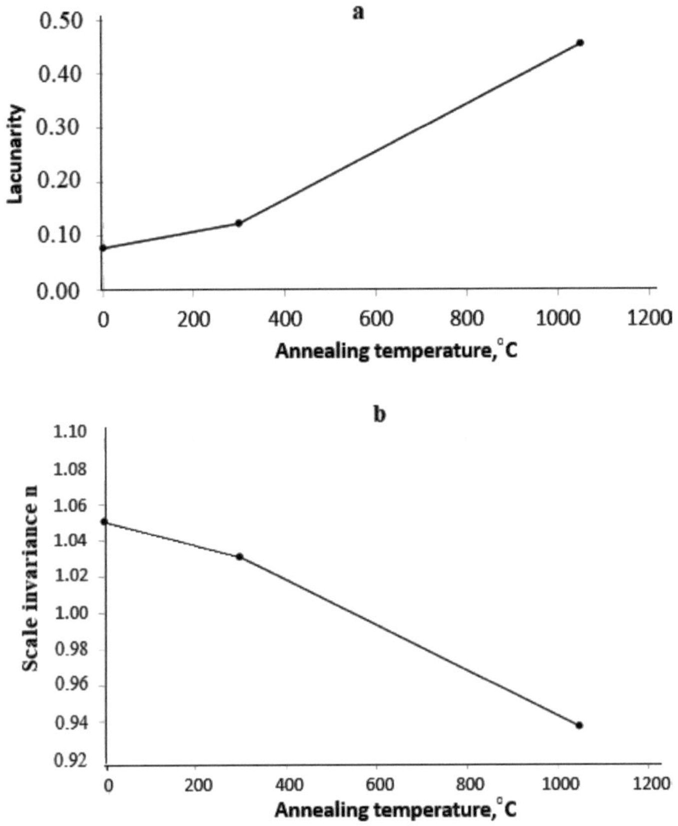

Figure 18.
Lacunarity (a) and scale invariance factor (b) as a function of the annealing temperature.

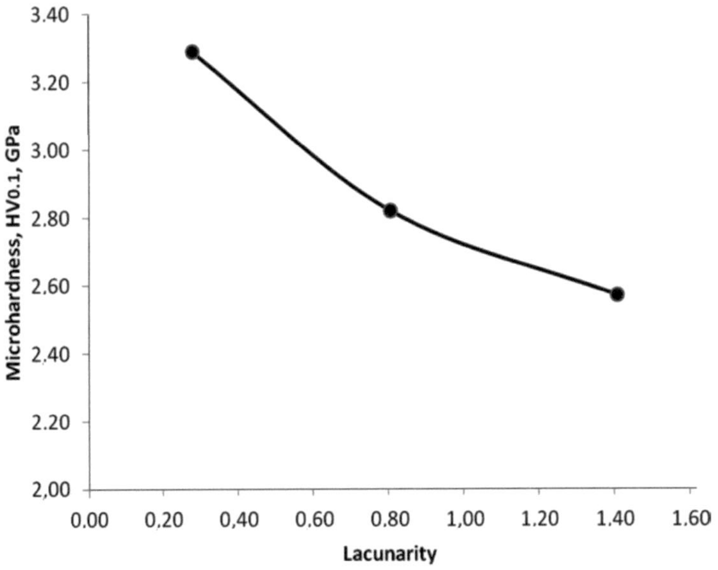

Figure 19.
Microhardness of steel samples as a function of lacunarity.

The relationship of microhardness with the uniformity of the mutual arrangement of nonmetallic inclusions was also analyzed. As shown in **Figure 19**, microhardness of the studied 3D-printed steel grows with the decrease of lacunarity, i.e., with the increase of no-metallic inclusions distribution homogeneity.

4. Conclusion

The presented examples demonstrate that such fractal parameters as lacunarity and its scale invariance characterizing the uniformity and self-similarity of the structure at different scale levels are essential factors prominently correlating with different properties of various materials. The described methods can be applied to a wide range of functional materials such as alloys, composites, and ceramics in order to describe their microstructural features on the level of mutual arrangement of their components and predict their properties. Particularly, common trends toward an improvement of target characteristics with a decrease of lacunarity and increase of scale invariance are revealed for different materials and their properties, including mechanical characteristics of steels and special ceramics, dielectric properties of polymer-inorganic composites, and ion transport in tungsten oxide-based electrochromic devices. The considered effects are determined by the uniformity of chemical composition and microstructure elements (fillers, phases, inclusions, etc.). The analysis of correlations between the chemical composition of materials, their fabrication and processing methods, microstructural characteristics (involving the above fractal parameters) and properties is an essential goal of Digital Materials Science providing an exquisite approach to the selection of optimal synthetic procedures, prediction, and enhancement of their target characteristics.

Acknowledgements

The study was supported by a grant from the Russian Science Foundation (project no. 21-73-30019). The equipment of the Center for Collective Use "Composition, structure, properties of structural and functional materials" of the National Research Center "Kurchatov Institute"—Central Research Institute of Structural |Materials "Prometey" was used with financial support from the Ministry of Education and Science of the Russian Federation under the agreement No. 13.TsKP.21.0014 (unique identifier RF 2296.61321X0014).

Author details

Maxim Sychov[1,2], Andrey Chekuryaev[1,3] and Sergey Mjakin[1*]

1 Saint-Petersburg State Institute of Technology, Saint-Petersburg, Russia

2 Institute of Silicate Chemistry of the Russian Academy of Sciences, Saint-Petersburg, Russia

3 National Research Center "Kurchatov Institute"—Central Research Institute of Structural Materials "Prometey", Saint-Petersburg, Russia

*Address all correspondence to: sergey_mjakin@mail.ru

References

[1] Kimmig J, Zechel S, Schubert US. Digital transformation in materials science: A paradigm change in Material's development. Advanced Materials. 2021; **33**:2004940. DOI: 10.1002/adma.202004940

[2] Sychov MM, Chekuryaev AG, Bogdanov SP, Kuznetsov PA. Digital Materials Science. In: Khakhomov S, Semchenko I, Demidenko O, Kovalenko D, editors. Research in Science & Technological Education, Inter-Academia 2021, LNNS. Vol. 422. Singapore: Springer, Singapore; 2021. DOI: 10.1007/978-981-19-0379-3_15

[3] Shevchenko VY, Ilyushin GD. Cluster self-organization of intermetallic systems: New two-layer cluster-precursors $K57$ = Li@15(Ga_6Cu_9)@41($Cu_{15}Mg_{26}$) and $K41$ = 0@8(Mg_2Ga_6)@33($Li_6Mg_3Ga_{24}$) in the $Li_{10}Mg_{34}Cu_{24}Ga_{71}$-hP139 crystal structure and $K5$ = 0@Ca_2LiInGe in the $Ca_2LiInGe_2$-oP24 crystal structure. Glass Physics and Chemistry. 2023;**49**:3-14

[4] Shevchenko VY, Ilyushin GD. Self-organization of intermetallic systems: New cluster-precursors $K12$ = 0@12(Li_9Ge_3) and $K9$ = 0@9(Li_8Ge) $Li_{68}Ge_{16}$-*oC*84 in the crystal structure, $K11$ = 0@11(Li_6Ge_5) and $K6$ = 0@6(Li_5Ge) cluster-precursors in the $Li_{44}Ge_{24}$-*oC*68 crystal structure and $K6$ = 0@6(Li_3Ge_3) cluster-precursors in the $Li_{12}Ge_{12}$-*tI*24 crystal structure. Glass Physics and Chemistry. 2023;**49**:15-26

[5] Shevchenko VY, Makogon AI, Sychov MM. Modeling of reaction-diffusion processes of synthesis of materials with regular (periodic) microstructure. Open Ceramics. 2021;**6**: 1-4. DOI: 10.1016/j.oceram.2021.100088

[6] Arsent'ev M, Sysoev E, Makogon A, Balabanov S, Sychov M, Hammouri M, et al. High throughput screening of 3D printed architected materials inspired by crystal lattices: Procedure, challenges and mechanical properties. ACS Omega. 2023;**8**(28):24865-24874. DOI: 10.1021/acsomega.3c00874

[7] Shevchenko V, Balabanov S, Sychov M, Karimova L. Prediction of cellular structures mechanical properties with the geometry of triply periodic minimal surfaces (TPMS). ACS Omega. 2023;**8**(30):26895-26905. DOI: 10.1021/acsomega.3c01631

[8] Chekuryaev AG, Sychev MM, Myakin SV. Analysis of the structure of composite systems by means of fractal characteristics using the $BaTiO_3$–Fullerenol–CEPA system as an example. Physics of the Solid State. 2021;**63**: 858-864

[9] Sitek W, Trzaska J, Dobrzanski LA. Evaluation of chemical composition effect on materials properties using AI methods. Journal of Achievements of Materials and Manufacturing Engineering. 2007;**20**:379-382

[10] Guo H, Wang Q, Urban A, Artrith N. Artificial intelligence-aided mapping of the structure–composition–conductivity relationships of glass–ceramic lithium thiophosphate electrolytes. Chemistry of Materials. 2022;**34**(15):6702-6712. DOI: 10.1021/acs.chemmater.2c00267

[11] Deng Q, Lin B. Automated machine learning structure-composition-property relationships of perovskite materials for energy conversion and storage. Energy Materials. 2021;**1**:100006. DOI: 10.20517/energymater.2021.10

[12] Thoppil GS, Nie J-F, Alankar A. Hierarchical machine learning based

structure–property correlations for as–cast complex concentrated alloys. Computational Materials Science. 2023; **216**:111855. DOI: 10.1016/j.commatsci.2022.111855

[13] Adisorn T, Tholen L, Götz T. Towards a digital product passport fit for contributing to a circular economy. Energies. 2021;**14**(8):2289. DOI: 10.3390/en14082289

[14] Kockmann N. Digital methods and tools for chemical equipment and plants. Reaction Chemistry & Engineering. 2019;**4**:1522-1529. DOI: 10.1039/C9RE00017H

[15] Impact of Software on Chemical Engineering Design Processes: Advances and Challenges. Available from: https://www.collegenp.com/article/impact-of-software-on-chemical-engineering-design-processes/

[16] Balabanov SV, Makogon AI, Sychov MM, Evstratov AA, Regazzi A, Lopez-Guesta JM. 3D printing and mechanical properties of polyamide products with Schwartz primitive topology. Technical Physics. 2020;**65**(2):211-215. DOI: 10.1134/S1063784220020036

[17] Bogdanov S, Sychov M, Lebedev L, Mjakin S, Gravit M. Core-shell powders for additive manufacturing of articles for underground construction. Procedia Engineering. 2016;**165**:1579-1586. DOI: 10.1016/j.proeng.2016.11.896

[18] Kantaros A, Ganetsos T, Petrescu FIT. Three-dimensional printing and 3D scanning: Emerging technologies exhibiting high potential in the field of cultural heritage. Applied Sciences. 2023;**13**(8):4777. DOI: 10.3390/app13084777

[19] Gonçalves RSBP, Haueisen J. Three-dimensional immersion scanning technique: A scalable low-cost solution for 3D scanning using water-based fluid. Sensors. 2023;**23**(6):3214. DOI: 10.3390/s23063214

[20] Cobos SF, Norley CJ, Pollmann SI, Holdsworth DW. Cost-effective micro-CT system for non-destructive testing of titanium 3D printed medical components. PLoS One. 2022;**17**(10):e0275732. DOI: 10.1371/journal.pone.0275732

[21] Li P, António Rodrigues Pereira P, Navas H, editors. Quality Control—Intelligent Manufacturing, Robust Design and Charts [Internet]. London, UK: IntechOpen; 2021. DOI: 10.5772/intechopen.87736

[22] Khan J, Teli J, Teli SN, Hada BP. Reduction of Cost of Quality by Using Robust Design: A Research Methodology. International Journal of Mechanical and Industrial Technology. 2015;**2**:122-128

[23] Niu X, Qin S, Zhang H, Wang M, Wong R. Exploring product design quality control and assurance under both traditional and crowdsourcing-based design environments. Advances in Mechanical Engineering. 2018;**10**(12):168781401881439. DOI: 10.1177/1687814018814395

[24] Mu ZQ, Lung CW. Studies on the fractal dimension and fracture toughness of steel. Journal of Physics D Applied Physics. 2000;**21**(5):848. DOI: 10.1088/0022-3727/21/5/031

[25] Bouchaud E, Lapasset G, Planès J. Fractal dimension of fractured surfaces: A universal value? EPL (Europhysics Letters). 2007;**13**(1):73. DOI: 10.1209/0295-5075/13/1/013

[26] Bouchaud E. Fractal properties of fracture surfaces: Roughness indices and

relevant lengthscales. In: Materials Research Society Symposia Proceedings. USA: Materials Research Society; 2012. p. 367. DOI: 10.1557/PROC-367-83

[27] Fernández R, González-Doncel G, Garcés G. Fractal analysis of strain-induced microstructures in metals. Fractal Analysis—Selected Examples. London, UK: IntechOpen; 2020. DOI: 10.5772/intechopen.91456

[28] Tang HP, Wang JZ, Zhu JL, Ao QB, Wang JY, Yang BJ, et al. Fractal dimension of pore-structure of porous metal materials made by stainless steel powder. Powder Technology. 2012;**217**: 383-387. DOI: 10.1016/j.powtec.2011.10.053

[29] Alvarez AC, Passé-Coutrin N, Gaspard S. Determination of the textural characteristics of carbon samples using scanning electronic microscopy images: Comparison with mercury porosimetry data. Adsorption. 2013;**19**:2-4. DOI: 10.1007/s10450-013-9530-7

[30] Ling EJY, Servio P, Kietzig A-M. Fractal and lacunarity analyses: Quantitative characterization of hierarchical surface topographies. Microscopy and Microanalysis. 2016;**22**: 168-177. DOI: 10.1017/S1431927615015561

[31] Parbat D, Bhattacharjee U, Paria S, Das N, Chakraborty M. Fractal dimension and lacunarity based microscopic image texture characterization of coated and non-coated metallic substrates. Advances in Materials and Processing Technologies. 2022;**8**:2244-2258. DOI: 10.1080/2374068X.2021.1903724

[32] Plotnick RE, Gardner RH, O'Neill RV. Lacunarity indices as measures of landscape texture. Landscape Ecology. 1993;**8**:201-211

[33] Dong P. Lacunarity for spatial heterogeneity measurement in GIS. Geographic Information Sciences. 2000; **6**:20-25

[34] Mjakin SV, Sychov MM, Chekuryaev AG, Sudar NT. Adjustment of dielectric performances for polymer-inorganic composites by ferroelectric filler modification with graphene micro-additive. Materials Today Proceedings. 2020;**30**(3):603-605

[35] Chekuryaev AG, Sychov MM, Mjakin SV, Britov VP, Belyaev VV. Analysis of microstructure and fractal characteristics of polymer based dielectric composites with graphene-modified barium titanate filler by box-counting method. Glass Physics and Chemistry. 2022;**48**:659-663. DOI: 10.1134/S1087659622600892

[36] Ma Y, Luo H, Guo R, Zhou K, Zhang D. Enhanced performance in multilayer-structured nanocomposites using $BaTiO_3$ and $Ba_{0.8}Sr_{0.2}TiO_3$ decorated graphene hybrids. Ceramics International. 2018;**44**:20871-20876. DOI: 10.1016/j.ceramint.2018.08.092

[37] Nikolaev AF. Khimicheskaya technologia, svoistva I primenenie plastmass (Chemical Technology, Properties and Applications of Plastics). Leningrad: Khimia Publishers; 1977. 368 p (in Russian)

[38] Alekseev SA, Korsakov VG, Sychov MM, Likhacheva OV, Rodionov AG, Ezhenkova LL. Effect of donor-acceptor sites at a barium titanate surface on the properties of composites based on cyanoethyl polyvinyl alcohol. Russian Journal of Physical Chemistry A. 2006;**80**:601-604

[39] Sychov M, Nakanishi Y, Vasina E, Eruzin A, Mjakin S, Khamova T, et al. Core-shell approach to control acid-base properties of dielectric and permittivity of its composite. Chemistry Letters. 2015;**44**:97-199. DOI: 10.1246/cl.140926

[40] Sychov MM, Minakova TS. Kislotno-Osnovnye Kharakteristiki Poverhnosti Tverdyh Tel I Upravlenie Svoistvami Materialov I Kompozitov (Acid-Base Characteristics of the Surface of Solids and Adjustment of the Properties of Materials and Composites). Saint-Petersburg: Khimizdat Publishers; 2022. 288 p. ISBN 978-5-93808-382-0 (in Russian)

[41] Korsakov VG, Sychov MM, Mjakin SV. Fizicheskaya Khimija Tverdogo Tela (Physical Chemistry of Solids). Saint-Petersburg, Russia: Emperor Alexander I Petersburg State Transport University Publishers; 2008. 177 p (in Russian)

[42] Chekuryaev AG, Sychov MM, Perevislov SN, Ulanov VN. Digital characteristics of microstructure of diamond—Silicon carbide composites. Ceramics. 2023;**6**:1067-1077. DOI: 10.3390/ceramics6020063

[43] Weaire D, Rivier N. Soap, cells and statistics—Random patterns in two dimensions. Contemporary Physics. 1984;**25**:59-99

[44] Quint SB, Quintab SB, Pacholski C. Extraordinary long range order in self-healing non-close packed 2D arrays. Soft Matter. 2011;**7**:3735-3738

[45] Bormashenko E, Frenkel M, Vilk A, Legchenkova I, Fedorets AA, Aktaev NE, et al. Characterization of self-assembled 2D patterns with Voronoi entropy. Entropy. 2018;**20**:956. DOI: 10.3390/e20120956

[46] Semenova A, Eruzin A, Bezrukov P, Sychov M, Mjakin S. Materials Today: Proceedings. 2020;**30**:606. DOI: 10.1016/j.matpr.2020.01.415

[47] Matveev VA, Eruzin AA, Semenova AA, Mjakin SV, Katashev PA. Glass Physics and Chemistry. 2021;**47**(7, Suppl. 1):63. DOI: 10.1134/S1087659621070075

[48] Sychov MM, Chekuryaev AG, Bogdanov SP, Kuznetsov PA. Digital materials science: Numerical characterization of steel microstructure. In: Khakhomov S, Semchenko I, Demidenko O, Kovalenko D, editors. Inter-Academia 2021, Lecture Notes in Networks and Systems. Vol. 422. Singapore: Springer. pp. 159-169

Chapter 7

The Finite Fractal Distributions as Mean Grain Size Distributions of Granular Matter

Emőke Imre and Vijay Pal Singh

Abstract

The grading entropy is the statistical entropy of the finite discrete grain size distribution on N uniform statistical cells in terms of the N sieve cells, consisting of two terms, the base entropy and the entropy increment (depending on N), which have normalized forms as well (basically independent of N). Being the most adequate statistical variables, both physical phenomena and physical model parameters can be best described by their use. Among others, the normalized base entropy A can be used to measure internal stability, being related to erosion, piping and liquefaction phenomena. Its value classifies the grading curves. Each class (with a fixed value of A) has a mean grading curve with finite fractal distribution, the fractal dimension varies from minus to plus infinity. (These mean gradings indicate a unique relation between the four entropy coordinates and four central moments). The internally stable fractal dimensions - between 2 and 3 – are occurring in nature verifying the internal stability rule of grading entropy. The widespread fractal soils are formed by degradation, which has a directional grading entropy path, with different features in terms of non-normalized and normalized grading entropy coordinates.

Keywords: fractal dimension, internal stability, grain size distribution, granular matter, grading entropy

1. Introduction

The grading curves contain a large amount of data. Hence, definitions and rules based on a few specified particle diameters are used in practice. For example, considering the diameter values (d), the range of gravel is d = 2 to 64 mm, the range of sand is d = 0.06 to 2 mm and the range of silt is d = 0.002 to 0.06 mm. These statistical cell sizes indicate that the probability variable is the logarithm of d (since the sieve hole diameters are doubled).

The coefficient of uniformity ($C_U = d_{60}/d_{10}$) and the coefficient of curvature ($C_C = d_{30}^2/(d_{60}d_{10})$) are frequently used. A granular soil is considered well-graded if $C_U > 4$ and $1 < C_c < 3$, otherwise it is poorly graded. These descriptors can provide limited information and these approaches are not valid for gap-graded grain size distributions. The use of these is somewhat controversial, since the grading curve is generally represented in terms of the logarithm of *d*.

Statistically, the measured grading curve is a finite, discrete distribution curve with *N* uniform statistical cells in terms of the logarithm of diameter *d* (*N* sieves, **Figure 1**). The distributions are frequently gap-graded or bimodal, and, the four central moments in terms of the logarithm of diameter *d* are with limited use except the mean. Therefore, the most adequate statistical variables are the grading entropy coordinates.

The grading entropy is the statistical entropy of the finite discrete grain size distribution on *N* uniform statistical cells in terms of the *N* sieve cells, consisting of two terms, the base entropy and the entropy increment (depending on *N*), which have normalized forms as well (basically independent of *N*). Both physical phenomena and physical model parameters can be best described by their use. Moreover, the four central moments in terms of the logarithm of diameter *d* are with limited use except for the mean, since the distributions are frequently gap-graded or bimodal. The grading entropy coordinates ([1], see the history in the Appendix) can be used for gap-graded or bimodal mixtures, the represent of the grading curve (path) in the 2-dimensional space, to elaborate the rules e.g., for internal stability, segregation and filtering in the grading entropy diagram [1], to study degradation or soil maturing.

The content of the chapter is as follows: The normalized grading entropy coordinates can be used to classify the grain size distribution curves (on the basis of the internal stability rule and their degradation lifetime), to define a mean grading curve with fractal distribution in each class [2]. The mean grading curves can be used to clarify the mean relationships between the various statistics of the grading curves [3]. The grading entropy coordinates can be used to extend the present liquefaction, erosion and piping rules into a more general form. The grading entropy coordinates can be used to study the degradation or breakage process [4], and to represent the degradation of various rock material tests on the same diagram [5] where the internal stability rules are represented. The results of these tests indicate that there is a unique, material-independent entropy path at a given testing condition with directional properties. The natural soils are internally stable due to the entropy path of the degradation [6].

The grading entropy coordinates have many other uses (not treated here) e.g., in the explanation of dry density of granular matter, for soil classification and in the interpolation of functions between the model parameters and the grading curve, where the fractal grading curves have also importance in the experimental testing since these are mean grading curves.

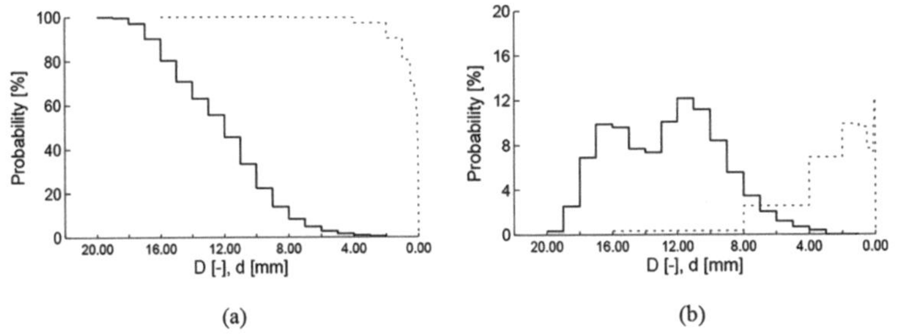

Figure 1.
(a, b): Grain size distribution function and density function in terms of diameter d (dashed line) and in terms of a kind of logarithm of d variable denoted by D (integer variable, solid line).

2. Grading entropy

2.1 Entropy of a statistical distribution

Let us consider M elements in m cells and, M_i elements in the i-th cell. The statistical entropy S_s is [7]:

$$S_s = Ms, \tag{1}$$

where s is the specific entropy or the entropy of an element given by

$$s = -\sum_{i=1}^{m} \alpha_i \log_b \alpha_i \tag{2}$$

In eq. (2), b is the base of the logarithm, and α_i is the relative frequency of the i-th cell, given by

$$\alpha_i = \frac{M_i}{M} \tag{3}$$

Let us consider two statistical cells, with relative frequencies of α_1 and α_2 = 1-α_1. If the base of the logarithm is set to 2 in Eq. (2) then the maximal specific entropy of this system is equal to 1 at the point $\alpha_1 = \alpha_2 = 0.5$ (**Figure 2(a)**). In the following, the base of the logarithm will be considered to be equal to 2. The specific entropy is written as follows, using $b = 2$, when the maximum ordinate is equal to 1:

$$s = -\frac{1}{\ln 2}\sum_{i=1}^{m} \alpha_i \ln \alpha_i \tag{4}$$

As it will be shown later on, this function – being similar to a half-ellipse – determines the shape of the boundary lines of the entropy diagram for small N (see Section 3.2).

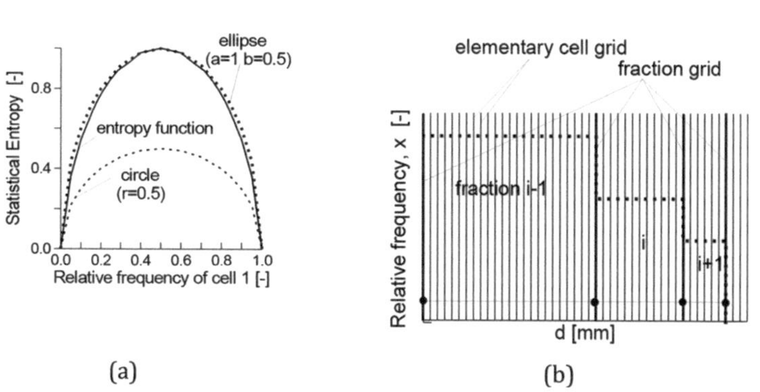

Figure 2.
(a) The entropy function of two cells, N = 2. (b) the fractions and the elementary cell grid, assuming uniform distribution within the fractions.

2.2 Statistical cells for the grading entropy

The statistical entropy of the grading curve is derived [1] from the definition of the entropy of a finite, discrete distribution [2], assuming a double cell system, the N "fractions" (uniform cell system in terms of logarithm of d) are embedded into the uniform elementary cell system in terms of d, defining a smallest diameter value d_0.

The statistical entropy formula is based on a uniform cell system. However, the measurement (using the sieve hole diameters) is made in a cell system uniform in the $\log d$ scale. Therefore, a double-cell system is used. The primary statistical cells or fractions are defined by successive multiplication with a factor of 2, starting from an (arbitrary) elementary cell width d_0 [8], the diameter range for fraction j:

$$2^{j+1}d_0 \geq d > 2^j d_0 \tag{5}$$

where d_0 is the elementary cell width (**Table 1**). Its possible value is equal to the height of the SiO_4 tetrahedron ($d_0 = 2^{-22}$ mm [9]). The number of the fractions N is the difference between the serial numbers of the finest and coarsest fractions as follows:

$$N = j_{\max} - j_{\min} + 1. \tag{6}$$

A secondary cell system is defined with equal width of d_0, assuming that the distribution is uniform within a fraction. The number of the elementary cells C_i in fraction i is equal to:

$$C_i = \frac{2^{i+1}d_{\min} - 2^i d_{\min}}{d_{\min}} = 2^i \tag{7}$$

The relative frequency of an elementary cell in fraction i is equal to:

$$\alpha_i = \frac{x_i}{C_i} \tag{8}$$

where x_i is the relative frequency of fraction i.

2.3 The definition of the grading entropy coordinates

The grading entropy S is the statistical entropy of the grading curve in terms of the elementary statistical cells (the statistical specific entropy of the grading curve in terms of the elementary cells). By inserting the relative frequency of the secondary cell α_i (i.e., Eq. (8)) into Eq. (4), the grading entropy S can be derived:

k	**1**	..	**23**	**24**
Limits	d_0 to $2\,d_0$	..	$2^{22}\,d_0$to $2^{23}\,d_0$	$2^{23}\,d_0$ to $2^{24}\,d_0$
S_{0k} or D [−]	1	..	23	24

Table 1.
Definition of fractions serial numbers and fraction entropies.

$$S = -\frac{1}{\ln 2}\sum_{k=i_1}^{i_N} x_k \ln x_k + \sum_{k=i_1}^{i_N} x_k \frac{\ln C_k}{\ln 2} S = \Delta S + S_o \quad (9)$$

It can be split into the base entropy S_0 and the entropy increment ΔS:

$$S_o = \sum_{i=1}^{N} x_i S_{oi} \text{ and } \Delta S = -\frac{1}{\ln 2}\sum_{i=1}^{N} x_i \ln x_i \quad (10)$$

$$S_{oi} = \frac{\ln C_i}{\ln 2}, S_{oi} = \frac{\ln 2^i}{\ln 2} = i \quad (11)$$

where S_{0k} (=D_k) is the grading entropy of the k-th fraction (**Table 1**), which is a positive integer, as it can be derived from Eqs. (7) and (10). It is uniquely related to logarithm of d, monotonically increasing with d (being about equal to the sieve or fraction serial number), expressing that the fraction cells are wider with d.

The base entropy S_0 is the mean fraction entropy weighed by the relative frequencies. The normalized base entropy, the so-called relative base entropy A:

$$A = \frac{S_o - S_{omin}}{S_{omax} - S_{omin}}, A = \frac{\sum_{i=1}^{N} x_i (S_{oi} - S_{o\,\min})}{N-1} = \frac{\sum_{i=1}^{N} x_i (i-1)}{N-1}. \quad (12)$$

The entropy increment ΔS is a smooth, concave function on the open simplex which may continuously be extended to the closed simplex. The normalized entropy increment B is defined as follows:

$$B = \frac{\Delta S}{\ln N} \quad (13)$$

The range of the entropy coordinates is dependent on N, ΔS varies between 0 and $\ln N/\ln 2$ and S_0 varies between S_{0min} and S_{0max}. The normalized coordinates are independent of the selection of d_0. The normalized coordinate B varies between 0 and $1/\ln 2$ and the relative base entropy A between 0 and 1.

3. The grading entropy map

3.1 The definitions

The spaces of the possible grading curves consist of such finite, discrete grain size distributions, which are composed of the same N consecutive fractions with minimum grain diameter d_{min} (being larger than or equal to d_0). The sum of the N relative frequencies x_i (i = 1, 2, 3...N) is equal to 1:

$$\sum_{j=1}^{N} x_j = 1, x_j \geq 0 \quad (14)$$

Eq. (14) is the equation of an N-1 dimensional closed simplex (**Figure 3(a)**), the N-1 dimensional analogy of the triangle or tetrahedron (i.e., the 2- and 3-dimensional

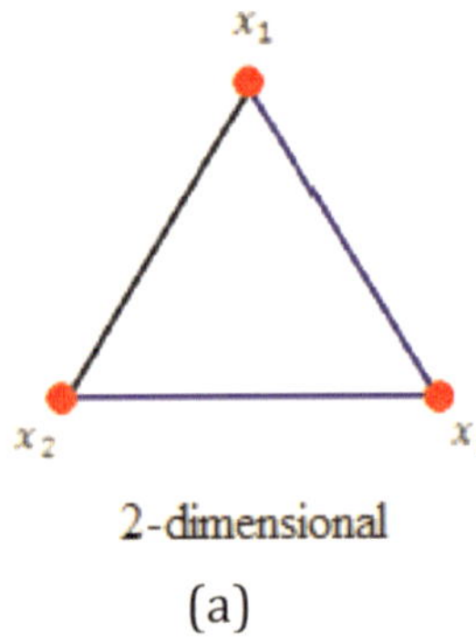

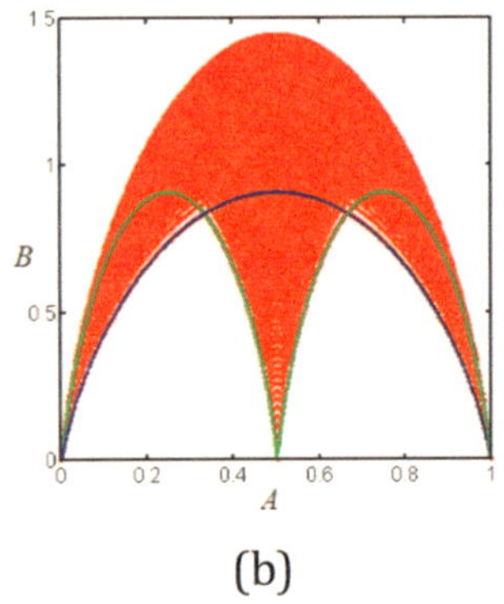

Figure 3.
Normalized entropy map, N = 2. (a) Standard simplex image with dimension 2 (after [10]). (b) the normalized grading entropy diagram for standard 2-dimensional simplex (after [10]).

instances) and the x_i (i = 1, 2, 3...N). The i-th barycentre coordinate (distance of the face generated by vertices 1 ... i-1, i + 1 ... N and the given simple point) is the same as the relative frequency x_i (i = 1, 2, 3...N). There is a one-to-one, isomorphic relationship between the space of the grading curves and the points of an N-1 dimensional closed simplex.

Four maps can be defined between a grading curve space (or an N-1 dimensional, open simplex with minimum grain diameter $d_{\min}$ (being larger than or equal to d_0)) and the 2-dimensional space of the entropy coordinates: the non-normalized $\Delta \rightarrow [S_0, \Delta S]$; normalized $\Delta \rightarrow [A,B]$; partly normalized $\Delta \rightarrow [A, \Delta S]$ or $\Delta \rightarrow [S_0, B]$. These are continuous on the open simplex and continuously extend to the closed simplex.

The diagram is compact, having a maximum and a minimum (normalized) entropy increment line which coincides with N = 2 (**Figure 3(b)**). In the case of the non-normalized map, every continuous subsimplex maps into the different parts of the diagram, following the structure. The normalized case entails nearly coinciding diagrams for every simplex with various N values, however, the upper and lower boundary lines are depending on N as described in the following.

3.2 The boundary lines of the entropy diagrams

The points of the maximum B line of the entropy diagram are the conditional maxima of B assuming the A = constant condition. The normalized entropy increment B is the logarithm of a generalized geometry mean of relative frequencies [11]. Being a symmetric and strictly concave function of the relative frequencies (with negative definite second derivative), it has a unique maximum for each A = constant value. The coordinates of the optimal point/grading curve [1]:

$$x_1 = \frac{1}{\sum_{j=1}^{N} a^{j-1}} = \frac{1-a}{1-a^N}, x_j = x_1 a^{j-1} \tag{15}$$

where a is the single positive root of the following equation:

$$y = \sum_{j=1}^{N} a^{j-1}[j - 1 - A(N-1)] = 0. \tag{16}$$

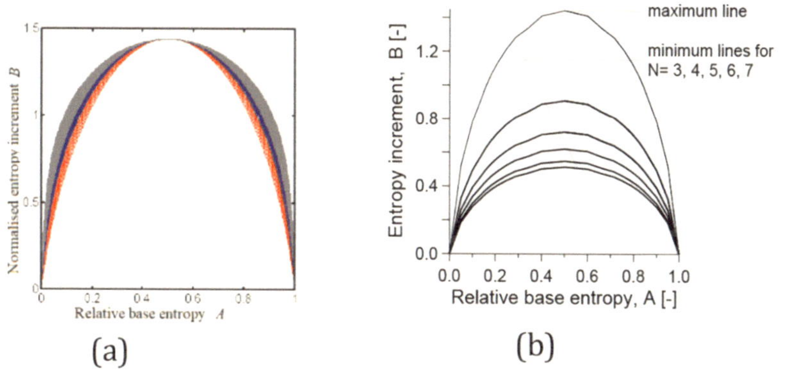

Figure 4.
(a) The maximum B lines, N varies between 2 and 100. (b) the approximate minimum B lines corresponding to the 1-N-edge of the simplex.

If $A < 0.5$ and the $A > 0.5$, then parameter a varies continuously between 0 to 1 and 1 to ∞, resp. The optimal points constitute a 1-dimensional line – the so-called optimal line – within the simplex for any N between vertices 1 and N, continuously depending on parameter A.

The optimal grading curves are changing continuously with parameter A, too, the optimal points are uniquely related to optimal grading curves. The optimal grading curves are unimodal, being either concave (if $A < 0.5$), linear if ($A = 0.5$) or convex if ($A > 0.5$), the relative frequencies x_i ($i = 1, 2, 3..N$) are decreasing, are constant and are increasing with decreasing diameter or fraction serial number, resp. The optimal grading curves have the shortest curve length graphs, among the grading curves with fixed A.

The points of the maximum B line of the entropy diagram slightly depend on N except at the symmetry point, where $a = 1$, $A = 0.5$ according to the following equation (**Figure 4(a)**):

$$A = \frac{1-a}{1-a^N}\frac{\sum_{i=1}^{N} a^{i-1}(i-1)}{N-1}, a \neq 1 \tag{17}$$

$$B = -\frac{1}{\ln N \ln 2}\left[\ln\frac{1-a}{1-a^N} + \ln a^{-\frac{a}{a-1}\frac{(N-1)a^N - Na^{N-1}+1}{(N-1)a^N - N+1}(N-1)}\right]. \tag{18}$$

The points of the minimum B line of the entropy diagram are the conditional minima of B assuming the A = constant condition. These points map from the vertices and the edges of the simplex (i.e., fractions, two-fraction mixtures (**Figure 4(b)**). The vertices 1, 2.. N have relative base entropy coordinate $A = 0, 1/(N-1), 2/(N-1) \dots (N-1)/(N-1) = 1$, respectively and zero B coordinate. The points of edge $i-j$ have the following A and B coordinates:

$$A = \frac{x_i(i-1) - x_j(j-1)}{N-1}. B = -\frac{1}{\ln N \ln 2}[x_i \ln x_i + (1-x_i)\ln(1-x_i)]. \tag{19}$$

In practice, the image of edge 1 – N is used as an approximate minimum B bounding line.

4. Properties of the entropy map

4.1 The inverse image

The inverse image of fixed A *and* B is the solution of the following system of equations:

$$f_1 = \sum_{i=1}^{N} x_i - 1 = 0, \tag{20}$$

$$f_2 = A - \frac{\sum_{i=1}^{N} x_i (i-1)}{N-1} = 0, \tag{21}$$

$$f_3 = B - -\frac{1}{\ln 2 \ln N} \sum_{x_i \neq 0} x_i \ln x_i = 0. \tag{22}$$

The regular values of the entropy map are the inner points of the entropy diagram. It follows from the regular value theorem [12, 13] that the inverse image of a regular value is a manifold, these are disjunct subspaces of the simplex. For $N > 2$, it is an N-3 dimensional manifold, which is "circle," situated on the A = constant, N-2 dimensional hyperplane section of the simplex, or in the isomorphic space of the grading curves (**Figure 4**).

The radius is determined by the difference $B_{max}(A)$ - $B(A)$. The critical values of the entropy map are the points of the maximum B line. The inverse image of a point of the maximum line of B *is* a single inner point of the simplex, called the optimal point, following from the concaveness of the normalized entropy increment B (**Figures** 5 and **6**).

The grading entropy parameter A is a linear function, the A = constant condition defines parallel hyper-plane sections of the N-1 dimensional simplex, which are related to the level lines of the A function (**Figure** 7). The grading entropy parameter

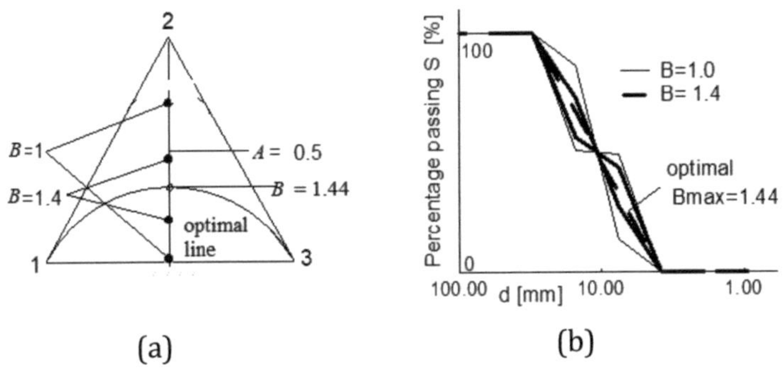

Figure 5.
Inverse image of fixed A, B, N = 3, 0-dimensional circles defined by 2 points. (a) in the simplex (b) in the space of grading curves.

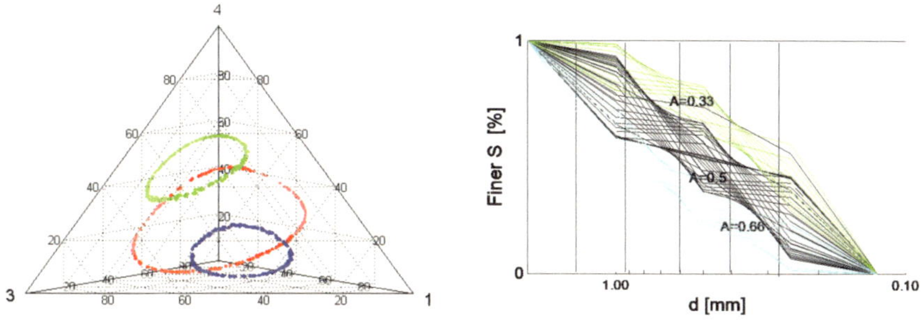

(a) (b)

Figure 6.
Inverse image of fixed A, B, N = 4, 1-dimensional circles (a) in the simplex (b) in the space of grading curves.

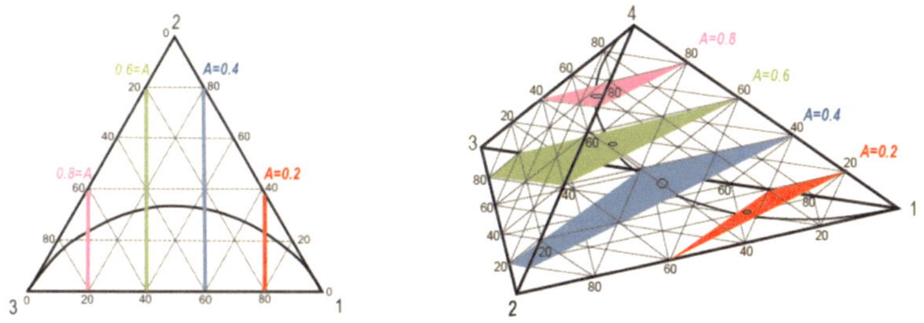

(a) (b)

Figure 7.
Inverse image of fixed A in simplex after [10], (a) and (b) N = 3 and 4. (It can be proved that the subgraph area in the grading curves is constant for a given A value).

B is a strictly concave function with a unique maximum for each A = constant value. The *B* = constant condition defines *N*-2 dimensional circles of the *N*-1 dimensional simplex which are related to the level lines of the *B* function (**Figure 8**).

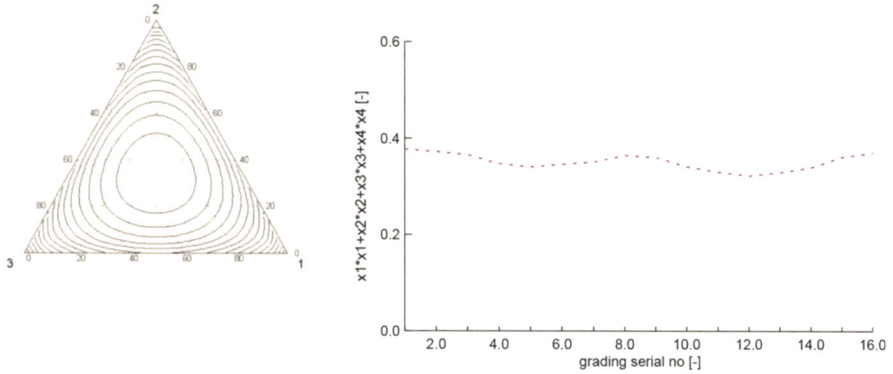

(a) (b)

Figure 8.
*(a) Inverse image of fixed B in the simplex, N = 3. (b) Sum of squared relative frequencies of the grading curves with A = 2/3, B = 1.2, N = 4 (**Table 2**).*

	Fractal	General grading curves with A = 2/3, and B = 1.2 (some examples)										
x_1 [-]	0.42	0.55	0.52	0.39	0.35	0.31	0.28	0.29	0.31	0.33	0.49	0.52
x_2 [-]	0.28	0.10	0.20	0.40	0.45	0.50	0.50	0.45	0.40	0.35	0.10	0.08
x_3 [-]	0.18	0.14	0.05	0.03	0.05	0.08	0.17	0.23	0.27	0.30	0.32	0.28
x_4 [-]	0.12	0.20	0.23	0.18	0.15	0.11	0.05	0.03	0.02	0.02	0.09	0.12
□x_1 [-]	0	0.13	0.1	−0.03	−0.07	−0.11	−0.14	−0.13	−0.11	−0.09	0.07	0.1
□x_2 [-]	0	−0.18	−0.08	0.12	0.17	0.22	0.22	0.17	0.12	0.07	−0.18	−0.2
□x_3 [-]	0	−0.03	−0.13	−0.15	−0.13	−0.1	−0.01	0.05	0.09	0.12	0.14	0.1
□x_4 [-]	0	0.08	0.11	0.06	0.03	−0.01	−0.07	−0.09	−0.1	−0.1	−0.03	0
$x_1{}^2$ [-]	0.1764	0	0.2704	0.1521	0.1225	0.0961	0.0784	0.0841	0.0961	0.1089	0.2401	0.2704
$x_2{}^2$ [-]	0.0784	0.01	0.04	0.16	0.2025	0.25	0.25	0.2025	0.16	0.1225	0.01	0.0064
$x_3{}^2$ [-]	0.0324	0.0225	0.0025	0.0009	0.0025	0.0064	0.0289	0.0529	0.0729	0.09	0.1024	0.0784
$x_4{}^2$ [-]	0.0144	0.04	0.0529	0.0324	0.0225	0.0121	0.0025	0.0009	0.0004	0.0004	0.0081	0.0144
Sum of squares $_i$		0.30	0.38	0.37	0.35	0.35	0.36	0.36	0.34	0.33	0.32	0.36
Sum positive delta $_i$		0.21	0.21	0.18	0.2	0.22	0.22	0.22	0.21	0.19	0.21	0.2

Table 2.
Numerical example. The A = 2/3, some grading curves with A = 2/3, and B = 1.2, positive delta is the sum of positive difference in the relative frequency with the fractal grading.

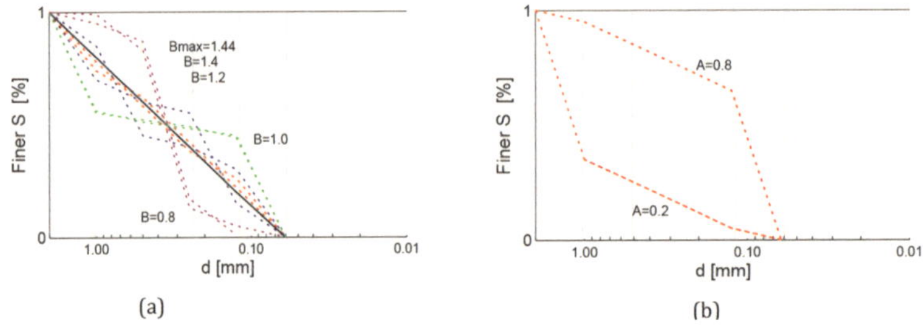

Figure 9.
(a) The mean or central position of the optimal grading curve related to fixed A = 0.5, Bmax = B = 1.44 in the class related to A = 0.5. (b) Symmetric grading curves.

4.2 Properties of the optimal grading curves

The A = constant condition defines a class of discrete distribution curves with identical subgraph area [2]. It follows from this that the optimal grading curve has a central position such that each grading curve has equal areas above and below the optimal in the same class of the gradings with A = constant.

The sum of the differences $(x_1 - x_{1\text{optimal}}) + \ldots + (x_N - x_{N\text{optimal}}) = 0$ is zero, the sum of the absolute differences (denoted by delta) is related to the difference between $B_{\max}$ of the optimal grading curve and the B of the actual distribution in the same class (**Figures 8(b)** and **9(a)**, **Table 2**).

The optimal grading curves (and all grading curves) are pair-wise symmetric (**Figure 9**). If the same relative frequencies x_i (i = 1 … N) are used in reverse order (e.g., $x'_1 = x_N$, $x'_2 = x_{N-1}$ ..) then A' = 1-A and B' = B. Two symmetric grading curves are shown in **Figure 9(b)**. The optimal grading curve with A and a is symmetric to the one with A' = 1-A and a' = 1/a, such that the same x_i (i = 1, 2, 3..N) occurs in reverse order (the symmetry for the optimal gradings appears by dividing polynomial in Eq. (16) with a^{N-1}).

4.3 The link of optimal and fractal grading curves

Let us examine the grading curves with N fractions and with finite fractal distribution which is described as follows [14, 15]:

$$F(d) = \frac{d^{(3-n)} - {d_{\min}}^{(3-n)}}{{d_{\max}}^{(3-n)} - {d_{\min}}^{(3-n)}} \tag{23}$$

where d is particle diameter, and n is fractal dimension. Taking into account that the fraction limits are defined by using the integer powers of the number 2, the relative frequencies of the fractions x_i (i = 1, 2, 3...N) can be expressed as follows:

$$2^j d_0 \geq d > 2^{j-1} d_0, \tag{24}$$

$$x_1 = \frac{\left(2^{(3-n)} - 1\right) {d_{\min}}^{(3-n)}}{{d_{\max}}^{(3-n)} - {d_{\min}}^{(3-n)}}, x_j = x_1 a^{j-1} \tag{25}$$

$$a = 2^{(3-n)}, n = 3 - \frac{\log a}{\log 2} \tag{26}$$

It follows that there is a one-to-one relation between the fractal dimension n and parameter a. The n varies between 3 and -∞ on the A > 0.5 side and between 3 and ∞ on the A < 0.5 side of the diagram (**Figure 10**), the optimal grading curves and only these have finite fractal distribution. The relation between n and A:

$$A = \frac{1 - 2^{3-n}}{1 - 2^{N(3-n)}} \frac{\sum_{i=1}^{N} (i-1) 2^{(i-1)(3-n)}}{N-1}, a \neq 1 \tag{27}$$

The parameter a and fractal dimension n are dependent on A and N (as shown in **Figure 10**) except at the symmetry point of the diagram with n = 3 and a = 1 irrespective of N. Some further examples are shown in **Tables 3–5** and **Figure 11**

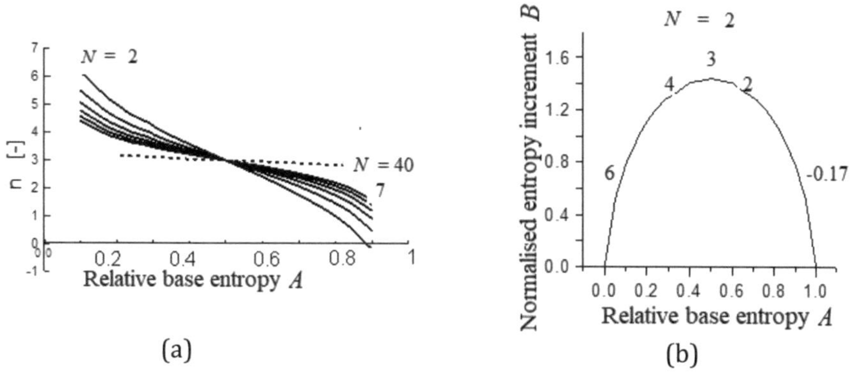

Figure 10.
(a) Fractal dimension n in terms of A and fraction number N. (b) N = 2 and 7, fractal dimension n in coordinates A = 0.1, 0.333, 0.5, 0.666, 0.9.

N [−]	2	3	4	5	6	7
A [−]	0.81	0.86	0.89	0.91	0.92	0.93

Table 3.
The A coordinates for a fixed fractal dimension n = 1 (and a = 4), various N.

N [−]	2	3	4	5	6	7
n [−]	2.00	2.25	2.40	2.49	2.56	2.62

Table 4.
The fractal dimension n in the function of N at fixed A = 2/3.

N [−]	2	3	4	5	6	7	8	15	30	40
A [−]	0.67	0.71	0.76	0,79	0.82	0,84	0.86	0.93	0.97	0.97

Table 5.
The A coordinates for a fixed fractal dimension n = 2 (and a = 2), various N for the optimal point at the global maximum of the grading entropy S.

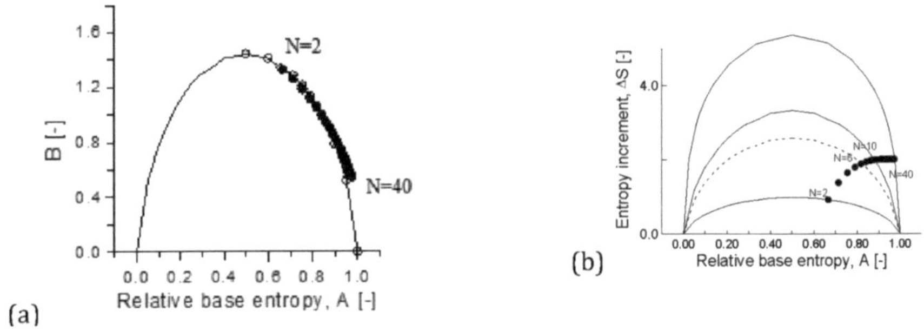

Figure 11.
The maximum grading entropy S points, N varies between 2 and 40. (a) the normalized entropy diagram. (b) Partly normalized diagram.

which illustrates how the points related to the maximum grading entropy S may vary with N, the connecting line has a role in the internal stability.

5. Internal stability rule

5.1 The internal stability rule for fractal soils, the probability of internal stability

The rule of the grading entropy theory [1, 16] was determined as follows. Simple vertical flow tests were designed and executed using artificial mixtures of natural sand grains. The dimensions of the permeameter were 20 cm in height and 10 cm in diameter. It was closed at the bottom by a sieve permeable of grains smaller than 1.2 mm but which retained grains larger than 1.2 mm.

The downward hydraulic gradient i was between 4 and 5. The two parts of the permeameter were separated after the test and the grading curves were determined, the grain movement was detected from these [1]. The results of the vertical water flow

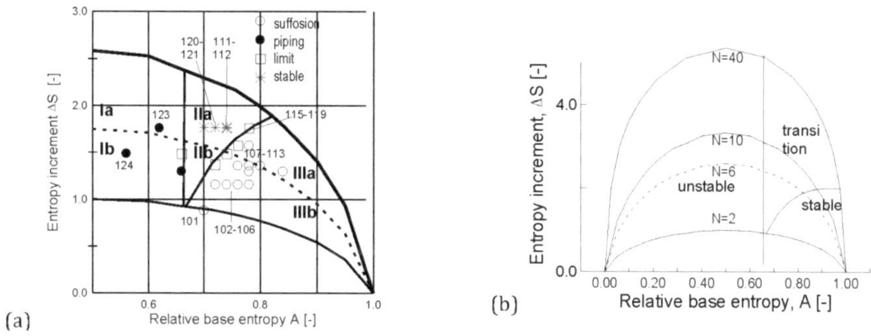

Figure 12.
Internal or grain structure stability criterion in the partly normalized diagrams. (a) Mixtures with N = 6 fractions (on the basis of [1]), with test outcome. (b) N = 6, 10, 40 fraction diagrams, simplified.

(suffosion) test were represented in the partly normalized entropy diagram, in terms of the relative base entropy and the entropy increment coordinates A and ΔS, as shown in **Figure 12(a)**.

According to the results, in Zone I (if $A < 2/3$) piping occurs which can be interpreted such that no structure of the large grains is present, the coarse particles "float" in the matrix of the fines and become destabilized when the fines are removed by piping. This zone is separated by the 2/3 vertical line. In the Zones II and III ($A = 2/3$ and $A < 2/3$), piping cannot occur, the coarse particles gradually form a skeleton, separated by the line connecting the maximum grading entropy S points, N varies (see **Figure 11**).

The gap-graded grading curves with N fraction – where more than 2 fraction is missing - are found at the lower part of the diagram (indicated by letter b), below the maximum entropy increment line related to $N - 2$, and in this case suffusion may occur. This follows from the Terzaghi's filter rule stating that there can be no more than two empty particle size fractions between the filter and the base soil:

$$1 \leq \frac{D_{\min}}{d_{\max}} \leq 4 \tag{28}$$

where D_{min} and d_{max} denote the diameter size of the filter and the diameter size of the base soil.

5.2 The internal stability rule for fractal soils, the probability of internally stable soils

As it was shown in Section 4.3, the fractal dimension n depends on N and A, except $A = 0.5$ as shown in **Figure 10**. **Table 6** combines the internal stability rule and the fractal dimension. The fractal soil is in the stable Zone III if $n < 2$, independently of the value of N. The transitionally stable Zone II is between $n = 2$ and the n related to $A = 2/3$ varying in the function of N.

The geometrical probability of being a fractal soil expressed by the ratio of the volume of the optimal line of the simplex and the volume of the whole simplex, is trivially equal to zero. The geometrical probability of internal stability can be expressed by the ratio of the volume of the simplex of the grading curves where $A > 2/3$ is met and the volume of the whole simplex which tends to zero as follows.

A [–]	$N = 2$	$N = 4$	$N = 5$	$N = 6$	$N = 7$
0.50	3.00	3.00	3.00	3.00	3.00
0.66	2.00	2.40	2.49	2.57	2.62
0.75	1.37	2.00	2.16	2.28	2.36
0.79	1.09	1.81	2.00	2.13	2.24
0.82	0.82	1.64	1.85	2.00	2.12
0.84	0.58	1.47	1.70	1.87	2.00

Table 6.
Fractal dimension and internal stability.

N [–]	**2**	**5**	**10**	**20**	**50**
P($A > 2/3$)	3E-01	1E-01	4E-02	6E-03	2E-05

Table 7.
Geometrical probability of stability in terms of the fraction number.

The ratio of the volume of the simplex (ie., grading curve space) where the internal stability condition is met (A > 2/3), versus the volume of the N-1 dimensional simplex) is the geometrical probability of an arbitrary grading curve with N fractions being internally stable. This probability is decreasing with the fraction number (e.g., for $N = 2$, it is equal to 1/3, for $N = 3$, it is equal to 2/9, for $N = 10$, it is equal to 4/100, for $N = 30$, it is equal to 8/10000, see **Table 7**). It decreases with N (Imre – Talata [6]) while most granular soils in nature are internally at least transitionally stable.

6. Examples for the verification of the internal stability rule

6.1 The internal stability criterion and piping problems

The Gouhou dam was a concrete-faced rock-fill dam 71 in high, directly founded on a sandy gravel of approximately 10 m thick [17]. The dam crest was 265 in long, and 7 m wide. The upstream and downstream slopes were 1:1.61 and 1:1.50, respectively. The damage was a piping breach. According to **Figure 13(a)**, by the testing the rock fill materials in the grading entropy diagram for internal stability, the soils had unstable internal structures. The riverbed was not unstable alone, and all grading curves had fractal-like distribution.

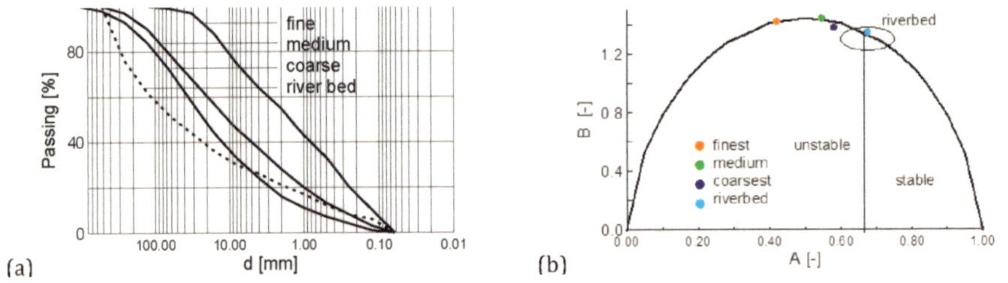

Figure 13.
Piping of Gouhou dam. (a) Gradings of soil for the river bed, the finest (I), medium (II) and coarsest (III) rockfill material. (b) Gradings in the entropy diagram.

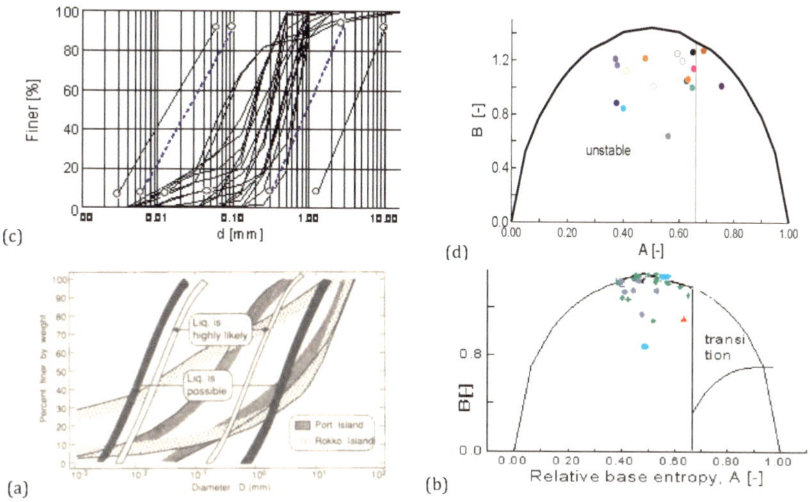

Figure 14.
(a) The liquefaction criterion for washed-out soil gradings at Hungarian piping river sites. (b) the piping soils in the entropy diagram [23, 24]. (c) the liquefaction criterion by ref. [18, 20–22] for Kobe grading curves. (d) the liquefied soils in the entropy diagram (light blue: Koebe soils).

The most frequent damage in the river dyke system of Hungary is piping [18, 19]. The grading curve of the washed-out soils is silty sand and fine sand with no cohesion, with d_{10} values within two orders of magnitude and with a small uniformity coefficient ($\sim C_u < 5$). Both the entropy-based and classical criteria of [20–22] prove the potential for these sands to liquefy (**Figure 14**). According to the result of an analysis, there is a combination of static and dynamic liquefactions leading to piping here [18, 19].

6.2 The internal stability criterion and liquefaction problems

Several seismic-induced liquefaction case studies were re-analyzed and tested for different susceptibly criteria on seismic-induced liquefaction, some results are shown in **Figure 14** [23, 25]. For example, the liquefied soils in the Kobe earthquake were internally unstable, while other criteria did not indicate problematic features [20–22].

Concerning static liquefaction, some triaxial tests series were made on Houstun sand mixed with various fine contents (**Figure 15**, [26]) indicating that the mixtures

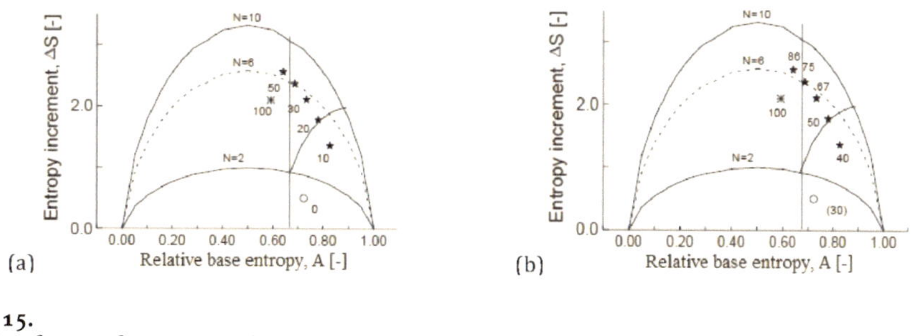

Figure 15.
Static liquefaction of Houstun sand – Fine mixtures. (a) Varying fine content in %. (b) Varying static liquefaction susceptibility in % [19, 25].

with increasing fine content and decreasing A coordinate the static liquefaction potential increased in the transitionally stable Zone II.

7. Fragmentation and degradation paths

The natural granular soils are generally fractal, with the stable fractal dimension between 2 and 3 ([26–38] (Section 5.2, 6) while the probability of being fractal or internally stable is zero or tends to zero with N (**Table 7**). This paradox can be resolved by the directional nature of the non-normalized entropy paths and a discontinuity of the normalized entropy coordinates path during degradation of relatively young mixtures (i.e., the minimum grain size or most grain sizes are larger than the few micron-sized comminution limit, see Kendall, [39]).

During fragmentation, the non-normalized entropy path is as follows: The base entropy S_0 decreases (since the mean diameter decreases), and the entropy increment ΔS monotonically increases (**Figure 16(a)**) (Stabilization with lime has the opposite effect in the entropy path [40]). The discontinuity of the normalized entropy coordinates path is described theoretically and experimentally in this section.

7.1 The theoretical entropy path at the appearance of new fractions

We assume that by definition, a possible space of the grading curves consists of such finite, discrete grain size distributions that are composed of the same N consecutive fractions with minimum grain diameter $d_{\min}$. Being larger than the crushing limit.

Keeping the value of N constant, the normalized and the non-normalized diagrams have the same structure, since the normalized and non-normalized coordinates are uniquely related. If we change the value of N by adding some zero fractions, the normalized coordinates of grading curve will differ (**Figure 17**). The difference in the case when i = 1,2 ... zero fractions are added from the smaller side for a PSD:

$$A(N+i) - A(N) = \frac{i[1 - A(N)]}{N + i - 1}) \tag{29}$$

$$B(N) - B(N+i) = \Delta S(N) \frac{1 - \frac{\ln N}{\ln N+i}}{\ln N} \tag{30}$$

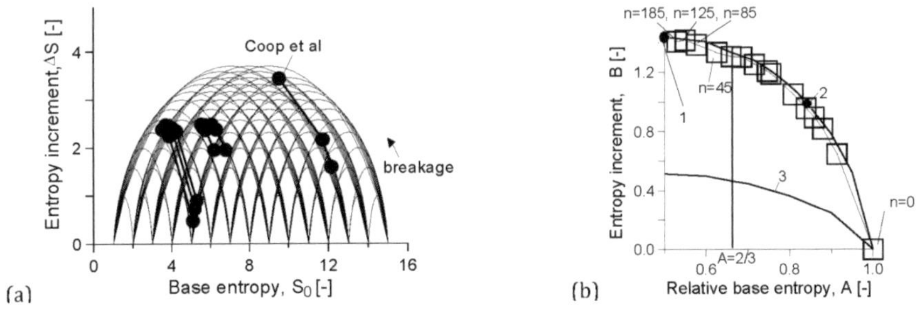

Figure 16.
(a) Non-normalized entropy paths for the laboratory tests treated in section 7. (b) Normalized entropy path of an initially one-fraction soil with the serial number of the crushing treatment, n. legend: 1: Maximum B point. 2: Maximum S point. 3: Approximate minimum B line [4].

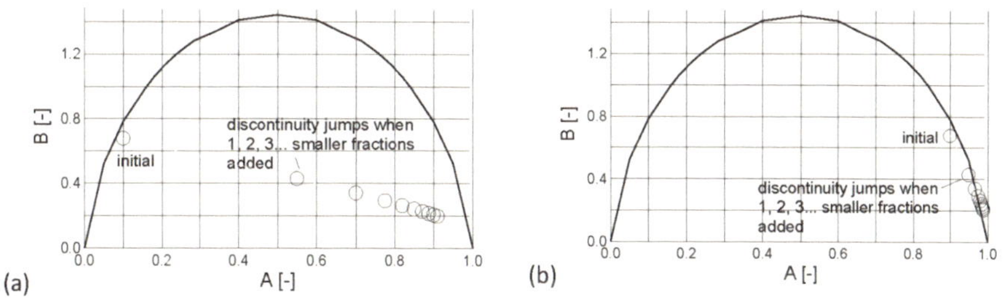

Figure 17.
The normalized entropy diagram, representing the effect of zero smaller fractions. (a) Initially N = 2, A = 0,1. (b) Initially N = 2, A = 0.19.

According to **Figure 17**, by adding 1,2..*i* zero fractions from the smaller side, the relative base entropy A will increase, which has some significance at fragmentation.

7.2 The entropy path in laboratory tests

The testing procedure developed to study the particle crushing phenomena was as follows [4, 5, 16]. The crushing treatments were made in a reinforced crushing pot, with the dimensions: diameter: 50 mm; height: 70 mm; wall thickness: 3 mm. Each treatment involved the application of a compressive load of 25,000 N to the sample contained in the crushing pot, using a loading machine. The sample was subjected to a series of crushing treatments (n = 1, 2 ... 10) then was removed from the crushing pot and the grading curve was determined by sieving. The sample was then returned.

Initially, single fraction quartz sand soil (**Figure 16(b)**) had an entropy path along the maximum B line, the velocity was decreasing close to the maximum entropy increment point. Theoretically it can be is important, that the fractal grading is ever present, though in case of other initial state, fractal grading would emerge later only [14, 15] as follows. Initially, two-fraction quartz and dolomitic sand mixtures with $A < 0.5$ [5] had a discontinuous entropy path at the start of the test due to the appearance of finer fractions, which drifted the entropy path into the stable part of the diagram (**Figure 18**). After the jump, a linear part entropy path occurred until the maximum entropy increment line was reached then the path followed it (**Figure 18**).

One-dimensional compression test was made with crushable grains (Light Expanded Clay Aggregate LECA) whose grains break at relatively low stress (see e.g.,

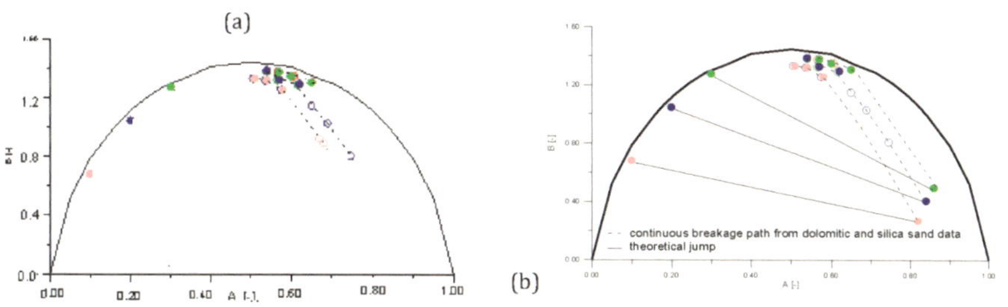

Figure 18.
Multi-compression tests with topology change and with 30 crushing treatments Trang et al. [5]. Normalized entropy path for initially N = 2 soils, every point indicates 10 crushing treatments, open symbol is quartz sand, closed symbol is dolomitic sand, the path is identical. (a) Measured data. (b) the computed jump in solid line and the subsequent path in dashed line, the entropy path is the same.

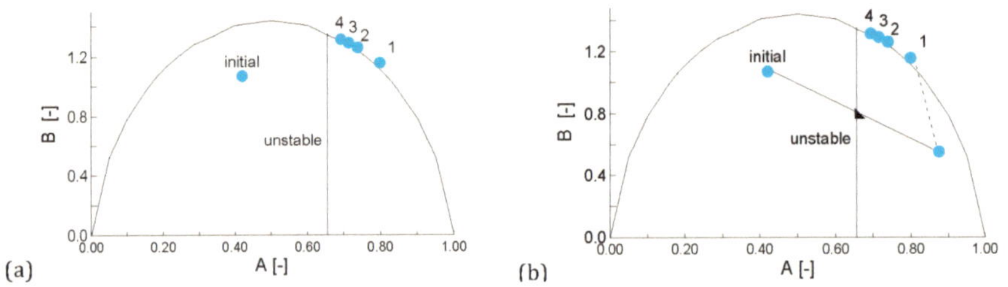

Figure 19.
Multi-compression test (Guida, parallel testing with the shear tests of Casini et al. [41–43]). (a) The normalized entropy path. (b) The computed jump in solid line the subsequent straight part in dashed line.

Casini et al. [41, 42]). The entropy path results are presented in **Figure 19** showing a similar pattern as the data in **Figure 18**.

7.2.1 Shear - compression type laboratory tests

Coop et al. [43] performed a series of ring shear tests to on a carbonate sand. Two test results are shown in **Figure 20(a)** and **Table 8**. One starts from a single-fraction, with similar results as **Figure 16(b)**. The other starts from a 3-fraction state on the $A > 0.5$ side of the normalized diagram, and after a jump where the A increased, a more stable final state with near fractal dimension between 2.4 and 2.8.

Dun [44, 45] performed multistage compression-shear tests on a hard rock (Andesite) and on a soft rock (Siltstone), the starting point was the same (**Figure 20(b)**).

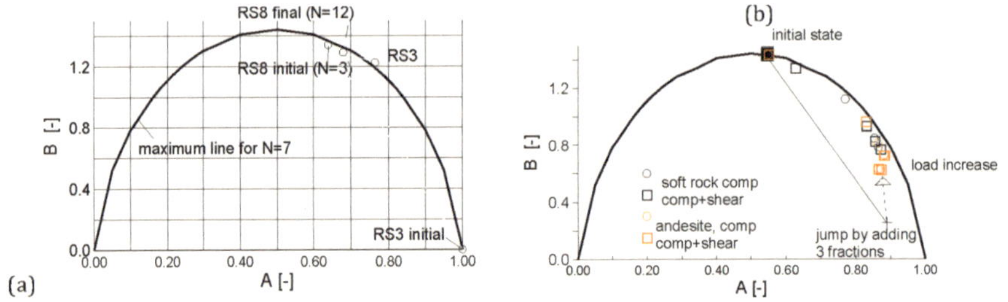

Figure 20.
Shear box test data (a) [43, 44], the initially single fraction and initially 3-fraction paths. (b) [18, 44], hard and soft rock, the computed jump in solid line the subsequent straight part in dashed line.

Sample	Rs3 (Initial grading 0,300–0,425)		Rs8 (Initial grading 0,063–0,425)	
State	**Initial**	**final**	**initial**	**final**
A [−]	1	0,744	0,640	0,66,884
B [−]	0	1224	1361	12,979
S_O [−]	13	10,18	12,28	9,36
ΔS [−]	0	3042	1495	3225
n [−]	- ∞	2,62	2,37	2,76

Table 8.
Shear box test data (N final = 12) in terms of entropy coordinates and the fractal dimension n of the closest optimal point.

Twenty five samples with particle sizes ranging between 19 mm and 4.75 mm were tested. The duplicate tests seemed to have the same normalized entropy paths for the two rocks.

7.3 *In situ* soil compaction, *in situ* maturity of natural soils, soil parameters in terms of grading curve

The results of an investigation of particle disintegration of natural sandy gravel and crushed stone bedding courses due to *in situ* compaction and the action of construction traffic in a dozen of different sites were re-evaluated in terms of grading curves before and after compaction [46]. According to the results, when the discontinuity formula was used to estimate the normalized coordinates after compaction, assuming Weibull distribution [42, 43] and fines, the normalized entropy path differed from the case when the smalls were neglected. The results presented in **Figure 21** show that with fines, the normalized parameters jumped into the transitionally stable region.

An example on natural degradation is the mine rehabilitation [47], where the concept of grading entropy was applied partly to evaluate the textural evolution of waste rocks at the surface of rehabilitated mine dumps, partly to clarify the internal stability change of the soil. A series of mine-spoil samples collected from topsoils and subsoils at different periods of time.

According to the results (**Figure 22**), the samples showed similar trends to the soils in many fragmentation problems. The base entropy S_0 decreased (since the mean diameter decreased), and the entropy increment ΔS monotonically increased in topsoil. However, unlike fragmentation by crushing, the largest fragments were not

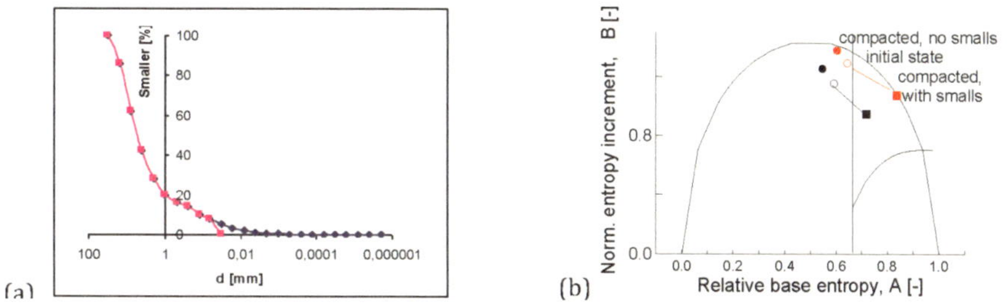

Figure 21.
Re-evaluation of compaction test results (after [46]). (a) the result of Weibull model fitting assuming 18 fractions instead of 10 fractions (resulting in about 0.06 difference in ΔS). (b) Initial and final states in the normalized diagram. Open symbol: Before, full symbol: After compaction. Circle: N = 10 and cube: N = 18 with completed fines.

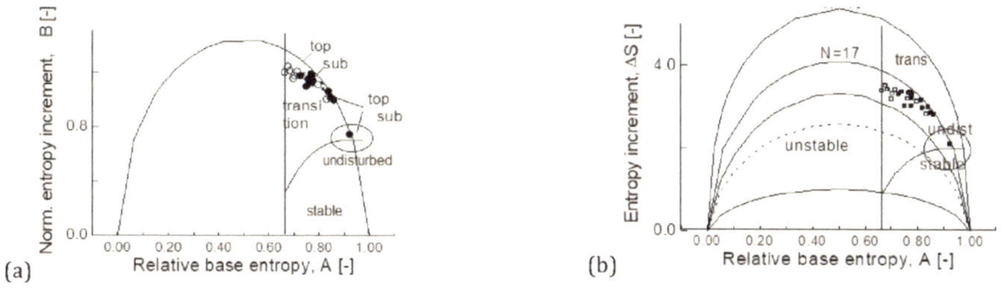

Figure 22.
Degradation of waste rock in open pit mine rehabilitation, the topsoil – Subsoil sample pairs are near fractal (the open/closed symbol: Topsoil layer/ subsoil [46]. (a) Normalized diagram (N = 17). (b) Partly normalized diagram. Topsoil is more degraded.

necessarily preserved [47]. The grading curves were with near fractal dimensions between 2.5 and 2.8. The samples plotted in the transitional stability zone, the most stable is the natural sub-soil.

The use of grading entropy in assessing granular physical properties (eg., soil hydraulic conductivity, dry density, soil water retention model parameters [48–51] is more and more accepted and the measurements related to the statistical relations are worthy to be made on fractal grading curves.

8. Discussion, conclusions

8.1 The entropy parameters, internal stability law

The grading entropy of a soil mixture is the statistical entropy based on uniform cells applied in terms of the log d (assuming uniform distribution within a statistical cell or on a sieve). It is the sum of two entropy coordinates: the base entropy S_0 and the entropy increment ΔS, with normalized versions (relative base entropy A, normalized entropy increment B) there are four entropy coordinates in this way.

The base entropy reflects the different sizes of the fraction cells and is a kind of mean log diameter, which is the suggested statistical variable of the grain size distribution, denoted by D. The relative base entropy A is a kind of normalized mean log diameter. Expressing the distance from the extremes (as a pseudo-metric), it has the potential to be a grain structure stability measure based on the simple physical fact that if enough large grains are present in a mixture, then these will form a stable skeleton.

The fully experimental internal stability rule of grading entropy [1] expresses that if $A >= 2/3$, then soil is internally stable and if $A < 2/3$, it is internally unstable. The experimental results indicate that internal stability depends continuously on A at least in the transitional stability zone (Section 6.2). Some DEM studies prove a clear link between internal structure and internal stability, the mechanical coordination number [2] was increasing in terms of A.

8.2 Soil classification on the basis of internal stability, mean grading curves

The normalized grading entropy coordinates – due to their such physical meaning that they express internal stability – may classify well the grading curves with N fractions. The grading curves with fixed A is a class with fixed internal stability, and the grading curves with fixed coordinate pair A and B is a disjunct subclass.

Each class has a unique optimal grading curve with maximum entropy increment, which is a fractal grading curve and a mean grading curve for the given class in the following sense. Every grading curve in the class has an equal subgraph area and deviates from the mean grading curve that equal areas are found above and below the mean grading curve, the difference is dependent on the B coordinate.

The optimal grading curves have a finite fractal distribution. The optimal or fractal grading curves have fractal dimensions between minus to plus infinity. The fractal soil is in the stable Zone III if $n < 2$, independently of the value of N. The transitionally stable Zone II is between $n = 2$ and the n related to $A = 2/3$ varying in the function of N.

The natural granular soils are generally fractal, with the stable fractal dimension between 2 and 3 [26–38] verifying the internal stability rule of grading entropy.

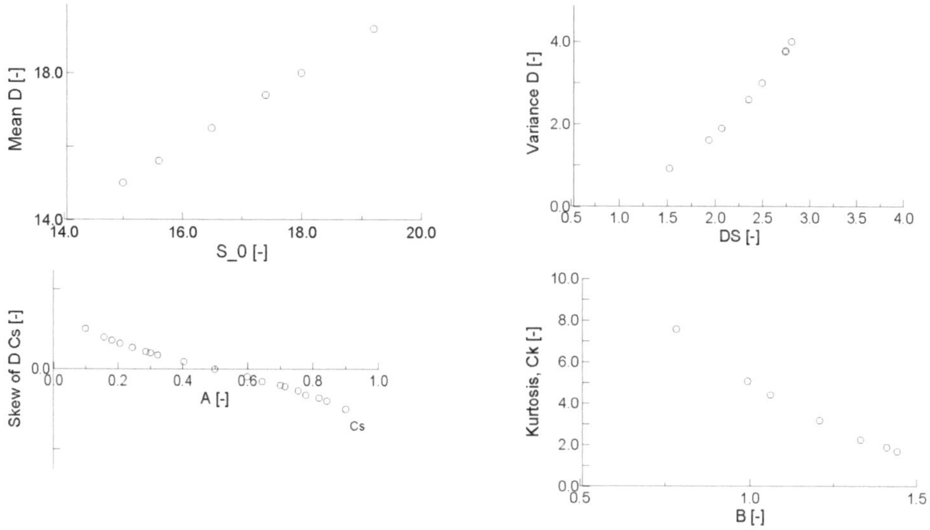

Figure 23.
Numerical experiment on the data of a series of mean, finite, discrete distribution functions with N = 7 statistical cells. (a) To (d): The relationships of the first four central moments in terms of the abstract diameter D [−] and the four entropy coordinates.

The mean functions constructed from simulated mean or fractal grading curves ([3], **Figure 23**) showed that entropy coordinates are similar to the four central moments as follows: The base entropy S_0 is identical to the first central moment and the entropy increment ΔS is similar to the variance, both are dependent on N. The relative base entropy A and the normalized entropy increment B are in one-to-one, smooth relation with the skew and the kurtosis, respectively, basically independently of N.

The entropy increment is better for the description of the dispersion than variance (i.e., second central moment) for gap-graded [46] and bimodal mixtures (see the Appendix). It measures how much the N fractions influence the soil behavior.

8.3 The entropy path and degradation

The space of the grading curves is isomorphic to an N-1 dimensional simplex, which is the continuous sub-simplex of a larger dimensional simplex. The grading entropy coordinates are smooth function on this simplex, the image of one base entropy and one entropy increment coordinates is called as entropy diagram. Every grading curve plot as a point in the diagram, a degradation process is a path in the diagram.

The natural granular soils are generally fractal, with a fractal dimension between 2 and 3 [26–38]. Since natural soils are generally within this fractal dimension bounds, the internal stability rule is verified by the nature. However, the probability of being fractal or internally stable is zero or tends to zero with N (**Table 7**). This paradox can be resolved by two features of the normalized entropy paths, the directional nature and a discontinuity feature of the degradation path.

During fragmentation, the non-normalized entropy path is the same, the base entropy S_0 decreases (since the mean diameter decreases), and the entropy increment ΔS monotonically increases (**Figure 16(a)**). (Stabilization with lime has the opposite effect in the entropy path [40]).

The fact that the space of the grading curves is isomorphic to an *N*-1 dimensional simplex, can be used to compute the geometrical probability that an accidental grading curve is fractal or internally stable which is zero and near zero decreasing with the fraction number *N* [6], respectively. This paradox that natural soils are mostly fractal and internally stable can be explained by the degradation process which is directional and the related non-normalized entropy path contains some discontinuity at the appearance of small fractions drifting the path into the stable domain as follows.

The extension of the grading entropy map onto a larger simplex is technically possible, by completing the original grading curves with zero fractions. The extended map is continuous for the non-normalized coordinates and has a discontinuity for the normalized coordinates. This fact has a practical significance since in nature the degradation results in some new fractions (the *N* increases), and the internal stability rule is formulated in terms of normalized or relative base entropy *A* which changes if *N* increases.

However, by changing the fraction number *N*, in the case of appearing some smaller fractions, a discontinuity occurs which can be computed. The computed path is a drift toward the more stable region. Then, after the discontinuity, the normalized entropy path under a constant fraction number is as follows. The relative base entropy *A* decreases and the normalized entropy increment *B* monotonically increases, until the maximum normalized entropy increment line is reached along a linear path. Then the path proceeds along it, through various fractal distributions with decreasing rates toward the symmetry points of the diagram. The path seems to be independent of the material, which may serve as a basis of some rock qualification or compaction control tests.

Acknowledgements

Thanks for the communications on particle breakage with Professor Itai Einav. The help of the Laboratories of the Department of Engineering Geology BME in the testing is acknowledged. This publication contains the work of PhD Student Phong Quoc Trang. Thanks for the contribution of dr. István Talata, dr. James Leak, dr. Daniel Barreto.

List of notations

A	relative base entropy
B	normalized entropy increment
C_U	uniformity coefficient given by the following expression d60/d10.
d_h	harmonic mean diameter
d_{10}	diameter of which 10% of the particles are finer
d_{60}	diameter of which 60% of the particles are finer
S_s	statistical entropy
S_0	base entropy
ΔS	entropy increment
M	Number of elements in statistical cell
M_i	Number of elements in the ith cell
m	Number of equivalent cells
s	Specific entropy

α	Frequency of elementary cells elements in the sieve cell
S	Total grading entropy
x	Relative frequency of PSD
N	is the number of fractions

A. Historical background (some pieces)

The Shannon entropy was first used for soils by Harrop-Williams, 1983 [52] using *d* as a statistical variable instead of log *d*. By a numerical example on gap-graded grading curves, the entropy was found to be better for the dispersion than variance. There is an example in **Figure A1** and **Table A1**. The variance is a maximum of the gap-graded grading curve, it is not maximal in terms of entropy, the gap-graded grading curves with *N* fractions plot below the *N*-2 maximum entropy increment line (see **Figure 12(a)**.

Lőrincz [1] elaborated the full theory, the relation with dry density and made the experiences for the grading entropy based rules. For the d_0, for the elementary cell width (**Table 1**), the height of the SiO_4 tetrahedron was suggested by Imre et al. ($d_0 = 2^{-22}$ mm [9, 53]). The first DEM applications were made eg. in [2], the first applications of the rules were made eg., by [9, 20, 25, 46, 52, 53].

The finite fractal distribution (by mass) formula was derived by Einav [14, 15]. The identity of the optimal grading curves of grading entropy (defined by Lőrincz [1]) and the finite fractal grading curves was revealed by Imre (published in Imre-Talata [6]). It was shown that the relations of the mean grading curves are mean relations (Imre et al. [3]).

Talata [40] revealed that the probability of internally stable mixtures decreases with *N* (Imre-Talata [6]). By using the simplex analogy [2], the probability of fractal mixtures is zero. In nature, the fractal gradings with fractal dimension 2 to 3 are very frequent (Section 6, [27–39]). The paradox is solved by the normalized grading entropy path [6]. The breakage process was related to grading entropy first by [1, 5]. The interpolation of model parameters in terms of the grading curve started by e.g., [1, 54–56].

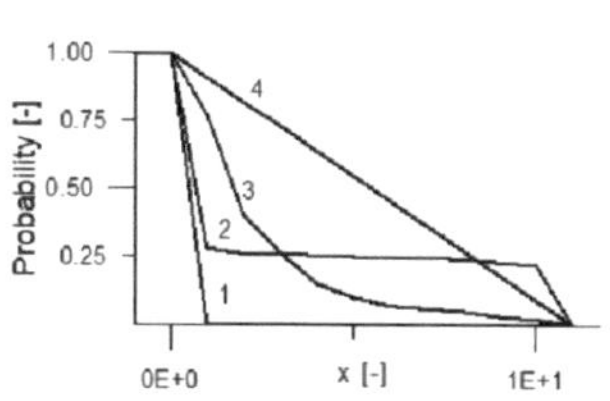

Figure A1.
Various distributions.

	distribution	1	2	3	4
mean	1	3.49	2.91	6	
variance	0	17.83	4.76	10	
entropy	0	1220	2559	3459	

Table A1.
Mean, variance and entropy increment of distributions on 11 cells, varying from 1 to 11.

Author details

Emőke Imre[1*] and Vijay Pal Singh[2]

1 EKIK HBM Systems Research Center and Bánki Donát Faculty of Mechanical and Safety Engineering, Óbuda University, Hungary

2 Texas A&M University, College Station, Texas, USA

*Address all correspondence to: imre.emoke@uni-obuda.hu

References

[1] Lőrincz J. Grading Entropy of Soils Doctoral Thesis. BME, Budapest (in Hungarian): Techn Sciences; 1986

[2] Imre E, Talata I, Barreto D, Datcheva M, Baille W, Georgiev I, et al. Some notes on granular mixtures with finite, discrete fractal distribution. Periodica Polytechnica. Civil Engineering. 2022;**66** (1):179-192

[3] Imre E, Hortobágyi ZS, Illés ZS, Nagy L, Talata I, Barreto D, et al. Statistical parameters and grading curves. In: Proceedings of the 20th International Conference on Soil Mechanics and Geotechnical Engineering. (20th ICSMGE). Sydney: Australian Geomechanics Society; 2022. pp. 713-718

[4] Lõrincz J, Imre E, Gálos M, Trang QP, Rajkai K, Fityus S, et al. Grading entropy variation due to soil crushing. International Journal of Geomechanics. 2005;**5**(4):311-319

[5] Trang PQ, Kárpáti L, Nyári I, Szendefy J, Imre E, L?rincz J. Entropy concept to explain particle breakage and soil improvement. In: Proc of the Young Geo Eng Conf. Alexandria, Egypt; 2009. pp. 87-90

[6] Imre E, Talata I. Some comments on fractal distributions and the grading entropy theory. In: Proceedings of the MAFIOK. Budapest: Szent István University; 2017. pp. 123-130

[7] Korn GA, Korn TM. Mathematical Handbook for Scientists and Engineers: Definitions, Theorems, and Formulas for Reference and Review. Courier Corporation; 2000

[8] Imre E. Characterization of dispersive and piping soils. In: Proceedings of the European Conference on Soil Mechanics and Foundation Engineering. (XI. ECSMFE), Copenhagen, 1995. 2. 49-55

[9] Imre E, Lõrincz J, Rózsa P. Characterization of some sand mixtures. In: Proc. of 12th IACMAG; October 2008. Goa, India; 2008. pp. 2064-2075

[10] Singh VP. Entropy Theory in Hydraulic Engineering: An Introduction. American Society of Civil Engineers. ASCE; 2014. p. 656

[11] Imre E, Kabey D. An extension of Thales' theorem-based geometric interpretation of the arithmetic-geometric means. Eng. Symposium in Bánki. 2022. pp. 157-163. Available from: https://bgk.uni-obuda.hu/esb/in_Hungarian.

[12] Milnor J, Weaver DW. Topology from the Differentiable Viewpoint. Princeton university press; 1997

[13] Hirsch MW. Differential Topology. Graduate Texts in Mathematics 1973 GTM, Volume 33 Hardcover ISBN: 978–0–387–90148-0 Copyright Information. New York: Springer-Verlag

[14] Einav I. Breakage mechanics—Part I: Theory. Journal of the Mechanics and Physics of Solids. 2007;**55**(6):1274-1297

[15] Einav I. Breakage mechanics—Part II: Modelling granular materials. Journal of the Mechanics and Physics of Solids. 2007;**55**(6):1298-1320

[16] Lőrincz J, Imre E, Fityus S, Trang PQ, Tarnai T, Talata I, et al. The grading entropy-based criteria for structural stability of granular materials and filters. Entropy. 2015;**17**(5):2781-2811. 31 p

[17] Zhang LM, Chen Q. Seepage failure mechanism of the Gouhou rockfill dam during reservoir water infiltration. Soils and Foundations. 2006;**46**(5):557-568

[18] Tsuchida H. Prediction and counter measure against the liquefaction. Seminar in the Port and Harbor Research Institution. 1970;**3**(1-3):33

[19] Rahemi N. Evaluation of Liquefaction Behavior of Sandy Soils Using Critical State Soil Mechanics and Instability Concept. PhD Thesis. Civil and Environmental Engineering Ruhr-University Bochum; 2017

[20] Barreto, Gonzalez D, Leak J, Dimitriadi V, McDougall J, Imre E, et al. Grading entropy coordinates and criteria for evaluation of liquefaction potential. In: Proceedings of the 7th International Conference on Earthquake Geotechnical Engineering. London: CRC Press; 2019. pp. 1346-1353

[21] Ishihara K. Terzaghi oration: Geotechnical aspects of 1995 Kobe earthquake. 1999. 14th ICSMFE

[22] Numata A, Mori S. Limits in the gradation curves of liquefiable soils. In: World Conf Earthquake Eng Vancouver, No. 1190. 2004

[23] Nagy L. The grading entropy of piping soils, Zbornik Radova Gradevinskog Fakukteta 20. In: Universitet u Novum Sadu Gradevinski Fakultet. Subotica; 2011. pp. 33-46. ISSN 0352-6852

[24] Imre E, Lorincz J, Trang PQ, Barreto D, Goudarzy M, Rahemi N, et al. A note on seismic induced liquefaction. In: XVII ECSMGE - Reykjavík, Iceland 1–6 of Sept 2019. IGS; 2019. pp. 979-988

[25] Imre E, Nagy L, Lorincz J, Rahemi N, Schanz T, Singh VP, et al. Some comments on the entropy-based criteria for piping. Entropy. 2015;**17**(4):2281-2303. 23 p

[26] Palmer AC, Sanderson TJO. Fractal crushing of ice and brittle solids. Proceedings of the Royal Society of London. Series A, Mathematical and Physical Sciences. 1991;**433**(1889):469-477

[27] Steacy SJ, Sammis CG. An automaton for fractal patterns of fragmentation. Nature. 1991;**353**:250-252

[28] Roberts JE, de Souza JM. The compressibility of sands. Proceedings of ASTM. 1958;**58**:1269-1277

[29] Hagerty MM, Hite DR, Ulrich CR, Hagerty DJ. One-dimensional high-pressure compression of granular material. Journal of Geotechnical Engineering. 1991;**119**(1):1-18

[30] Okada Y, Sassa K, Fukuoka H. Excess pore pressure and grain crushing of sands by means of undrained and naturally drained ring-shear tests. Engineering in Geology. 2004;**75**(3–4):325-334

[31] Sadrekarimi A, Olson SM. Particle damage observed in ring shear tests on sands. Canadian Geotechnical Journal. 2010;**47**(5):497-515

[32] Ho TYK, Jardine RJ, Anh-Minh N. Large displacement interface shear between steel and granular media. Geotechnique. 2011;**61**(3):221-234

[33] Pál G, Jánosi Z, Kun F, Main IG. Fragmentation and shear band formation by slow compression of brittle porous media. Physical Review E. 2016;**94**:053003

[34] Kun F. Breakage of particles, session introduction. In: P&G Conference, Montpellier, 2017. 2017

[35] Sammis CG, King G, Biegel R. Kinematics of gouge deformations pure. Applied Geophysics. 1987;**125**:777-812

[36] Airey DW, Kelly RB. Interface Behaviours from Large Diameter Ring Shear Tests Proceedings of the Research Symposium on the Characterization and Behaviour of Interfaces. Atlanta, GA, USA; 2008. pp. 1-6

[37] Imre B, Laue J, Springman SM. Fractal fragmentation of rocks within sturzstroms: Insight derived from physical experiments within the ETH geotechnical drum centrifuge. Granular Matter. 2010;**12**(3):267-285

[38] McDowell GR, Bolton MD, Robertson D. The fractal crushing of granular materials. Journal of the Mechanics and Physics of Solids. 1996; **44**(12):2079-2101

[39] Kendall K. The impossibility of comminuting small particles by compression. Nature. 1978;**272**(5655): 710-711

[40] Harrop-Williams K. Some geotechnical applications of entropy. In: Proc. of the 4th ICASP. Vol. 2. 1983

[41] Casini F, Giulia M, Viggiani B, Springman SM. Breakage of an artificial crushable material under loading. Granular Matter. 2013;**15**:661-673. DOI: 10.1007/s10035-013-0432

[42] Guida G, Bartoli M, Casini F, Viggiani GMB. Weibull distribution to describe grading evolution of materials with crushable grains. Procedia Engineering. 2016;**158**:75-80

[43] Coop MR, Sorensen KK, Bodas Freitas KK, Georgoutsos G. Particle breakage during shearing of a carbonate sand. Geotechnique. 2004;**54**(3):157-163

[44] Dun R. The Importance of Particle Breakage for Shear Strength Measured in Direct Shear. Undergraduate honours thesis. The School of Engineering. The University of Newcastle (unpublished); 2014

[45] Fityus S, Imre E. The significance of relative density for particle damage in loaded and sheared gravels powders and grains 2017. In: 8th International Conference on Micromechanics on Granular Media. 2017. DOI: 10.1051/epjconf/201714007011. 4 p

[46] Imre E, Lõrincz J, Trang PQ, Casini F, Guida G, Fityus S, et al. Reanalysis of some in situ compaction test results. In: Proc of the XVII ECSMGE-2019. Reykjavík; 2019. Paper: 0989. DOI: 10.32075/17ECSMGE-2019-0989

[47] Imre E, Fityus S. The use of soil grading entropy as a measure of soil texture maturity and internal stability. In: Proceedings of the (XVI. DECGE). Macedonia; 2018. pp. 639-644

[48] Feng S, Vardanega PJ, Ibraim E, Widyatmoko I, Ojum C. Permeability assessment of some granular mixtures. Géotechnique. 2019;**69**(7):646-654. DOI: 10.1680/jgeot.17.T.039

[49] O'Kelly BC, Nogal M. Determination of soil permeability coefficient following an updated grading entropy method. Geotechnical Research. 2020;7(1):58-70, 13 p. DOI: 10.5772/intechopen.69167

[50] Imre E, Firgi T, Baille W, Datcheva M, Barreto D, Feng S, et al. Soil parameters in terms of entropy coordinates. In: E3S Web of Conference. Vol. 382. UNSAT; 2023. p. 25003. DOI: 10.1051/e3sconf/202338225003

[51] Imre E, Barreto D, Datcheva M, Singh VP, Baille W, Feng S, et al. Minimum dry density in terms of

grading entropy coordinates. In: Proceedings of the 17th Asian Regional Conference on Soil Mechanics and Geotechnical Engineering; 14-18 August 2023; 17th ARC, Astana, Kazakhstan. London, UK: CRC Press; 2023. pp. 402-406, 5 p. DOI: 10.1201/9781003299127-42

[52] Imre E, Lörincz J, Szendefy J, Trang PQ, Nagy L, Singh VP, et al. Case studies and benchmark examples for the use of grading entropy in Geotechnics. Entropy. 2012;**14**:1079-1102

[53] Talata I. A volume formula for medial sections of simplices. Discrete & Computational Geometry. 2003;**30**:343-353. DOI: 10.1007/s00454-003-0015-6

[54] Imre E, Lőrincz J, Trang QP, Fityus S, Pusztai J, Telekes G, et al. A general dry density law for sands. KSCE Journal of Civil Engineering. 2009;**13**(4):257-272

[55] Imre E, Rajkai K, Genovese E, Fityus S. The SWCC transfer functions of sands. In: Proceedings of 4th Asia-Pacific Conference on Unsaturated Soils. 2009. pp. 791-797, 7 p

[56] Imre E, Rajkai K, Genovese E, Jommi K. A SWCC transfer functions of sands. In: Proc. of 5th Int. Conf. on Unsaturated Soils. 2010. pp. 453-459, 7 p

Chapter 8

Placental Growth and Development Analyzed through 2D and 3D Fractals

Ruchit Shah, Carolyn Salafia, Theresa Girardi, Pratipaul Persaud, Jessica Brunner, Emily S. Barrett, Thomas O'Connor and Richard K. Miller

Abstract

Fractal geometry has many applications in physiology and anatomy, providing novel insights into the structure and function of biological systems and organs, including the placenta. The placenta is a vital fetal organ that is the means by which essential nutrients and oxygen are extracted from maternal blood and transferred to the developing fetus. Anatomically, the placenta is a highly intricate structure exhibiting self-similarity at different scales. The complex relationship between placental function in nutrient transfer and fetal growth follows the allometric metabolic scaling law. The variety of shapes of the placental chorionic plate, a 2D plane that contains the major chorionic arteries and veins, has been linked to measures of child health and neurodevelopmental outcomes. The microscopic arrangement of chorionic villi has also been demonstrated to have fractal properties that vary by gestational age and in different pathological conditions. Geographical Information Systems theory could be used to analyze the placental topography in the context of its surface vasculature. Measures of spatial autocorrelation can model placental growth and development over gestation. An ideal model would mark the timing, nature, and severity of gestational pathology modifying placental growth and, by extension, fetal development that leads to poor pregnancy outcomes.

Keywords: placenta, Voronoi cells, chorionic plate vasculature, placental villi, villus packing density, geographic information systems

1. Introduction

"Fractal" refers to a particular type of complex geometric shape that is distinct from the simple figures of classical or Euclidean geometry—the square, the circle, the sphere, and so on. A fractal can be used to quantify an irregularly shaped object or natural phenomena like coastlines and mountain ranges. Mandelbrot was the first to suggest that fractals could be an ideal tool in applied mathematics to model diverse phenomena—from physical objects to stock market behavior. Since its inception, the fractal concept has spawned a new system of geometry that has had significant implications for fields as diverse as physical chemistry, physiology, and fluid mechanics [1–3].

Fractals are "self-similar" at all scales, at least approximately, if not precisely. The similarities between levels are not necessarily obvious but they are present—providing order to chaos; for this reason, fractal analysis is used to generate quantitative measures of the complexity and self-similarity of biological structures and processes. This quantitative measure is termed a "fractal dimension," which is an index for characterizing fractal patterns or sets by quantifying as a ratio of the change in complexity to the change in scale [4]. Fractal dimensions have been used to characterize physiological signals such as heart rate variability [5, 6] or respiratory rate [7, 8]. Comparing fractal dimensions can provide insight into the complexity and adaptability of these signals and has been applied to other conditions as diverse as sleep apnea, cardiovascular disease, and aging [9, 10].

Fractal analysis has also been applied to biological structures, such as the dichotomous hierarchical branching network structure of blood vessels [11, 12] and the morphology of (individual) neurons [13]. By quantifying the fractal properties of these structures, researchers can better understand their function and also developmental processes across time.

Fractal analysis has been used to study topics as diverse as the distribution of plant species in ecosystems [14] and the spatial distribution of neurons in the brain [15]. The mathematical analysis of large-scale spatial distributions and patterning of ecosystems and even social networks have all provided evidence that shared governing principles link rules that underlie the organization and self-organization of diverse systems.

The placenta is a disc-shaped organ primarily composed of blood vessels, connective tissue, and a thin trophoblast epithelium; weighs about a pound at full term; and is approximately an inch thick and 9 inches in diameter. The placenta comprises a fetal part and a maternal part (**Figure 1**). The fetal part consists of two membranes—the amnion and the chorion, the umbilical cord, the chorionic plate vasculature, and the villous tree. The maternal part consists of the decidua and the intervillous space filled with maternal blood.

The placenta is the anatomic and physiologic interface between the fetal and maternal environments during pregnancy. The newborn is entirely composed of nutrients transferred from the maternal to the fetal circulation across the placenta. By extension, birth weight (BW) depends on placental function. It has been demonstrated that the placental and fetal masses are related by an allometric exponent. Deviations from this range can illuminate physiological aberrations during the pregnancy and can be indicators of deviations from normal fetal-placental development.

In contrast to most permanent viscera that have more fixed shapes, the placenta is amoeboid and responds to intrauterine exposures by modifying its gross and microscopic structure and vascularization [16–25]. Shape variations that mark altered placental vascularization include irregularities of chorionic surface area, shape, thickness, and vascular patterning and cord insertion [26]. As compared to the traditional crude measures of length, breadth, and average thickness, a comprehensive measurement of placental shape and surface vessels has been shown to provide a powerful reflection of the intrauterine environment that can inform on postnatal risks [12, 27]. Fractal geometry has already helped to characterize and differentiate the branching patterns of placental chorionic surface vasculature and villous distribution at

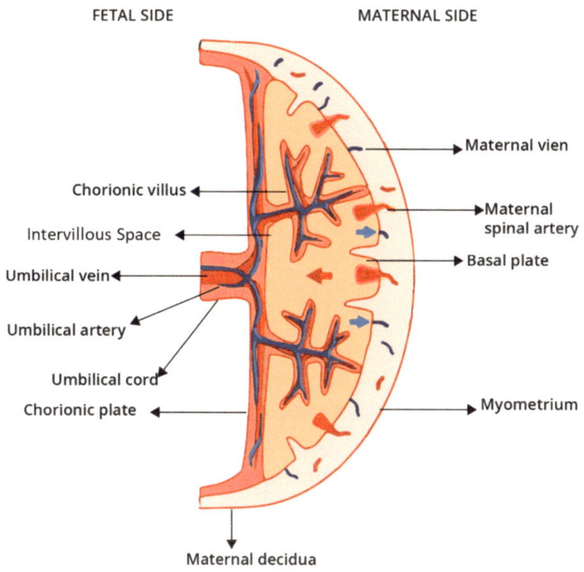

Figure 1.
Schematic diagram of a human placenta. The fetal side comprises chorionic plate, chorionic surface arteries, and veins connected to the umbilical arteries and vein. The maternal side consists of the maternal decidua and basal plate with maternal vessels. The middle portion comprises placental villi in maternal blood.

different gestational stages and for different pathological conditions [12, 28, 29]. Altered proportions of the chorionic plate (a very small chorionic plate area or a very large and expansive chorionic plate) or placental disk thickness (a thin or thick disk) has been correlated with changes in BW. Moreover, the changes in chorionic surface vascular resistance and extent of villous arborization would, in theory, represent very different fetal-placental relationships resulting from exposures to the intrauterine environment at different gestational time points and resulting in different lifelong health risks.

This chapter focuses on the anatomic and physiologic fractality of the placenta.

2. The fractal placenta

2.1 Placental weight

Allometric scaling, which describes how the size, shape, or physiology of an organ changes in relation to metabolism, had its first major proponent in Max Kleiber. Kleiber's law postulated that the metabolic rate of an organism scales as a power function of its body mass [30]. While he hypothesized that the value of the exponent should be 2/3, experimental data showed that ¾ was a better fit. The ¾ exponent was later justified using hydrodynamic optimality [31]. This scaling has proved remarkably constant for a wide range of organisms from the smallest microbes to the largest vertebrates and plants [32].

Ahern in 1966 (as cited in [33]) suggested that the complex relationship between placental function in nutrient transfer and fetal growth could also be analyzed allometrically by substituting placental weight (PW) for basal metabolic rate and fetal BW for body mass, as seen below.

$$Placental\ Weight = \alpha \times Birth\ Weight^{\hat{a}}$$

Ahern posited a 2/3 scaling for his relationship, as postulated by Kleiber. We tested this scaling relation in a sample of 24,601 term singleton pregnancies from the National Collaborative Perinatal Project [34]. The allometric metabolic equation was solved for α and β by rewriting $PW = \alpha(BW)^{\beta}$ as $\ln(PW) = \ln(\alpha) + \beta[\ln(BW)]$. The results obtained were in remarkable agreement with Kleiber's law: $\beta = 0.78 \pm 0.02$, $\alpha = 1.03 \pm 0.17$. Rearranging the terms in the equation, and since $\ln(1) = 0$, $\beta = [\ln(PW)/\ln(BW)]$. We interpret β as a non-linear measure of placental efficiency; when β increases, PW is greater relative to BW, implying reduced placental functional efficiency, and when β decreases, BW is relatively larger for a given PW (implying greater placental efficiency). While the fetoplacental weight ratio is more universal in clinical practice, it is a linear version of β, which contrasts with the logarithmic growth of the fetus and placenta approaching term. Thus, understanding the allometric scaling relationship between the placenta and fetus is important for clarifying the impact of environmental stressors on placental structure/function and the effects of that stressor on the fetus.

Another example of how broadly applicable this ¾ scaling relationship was found for twin pregnancies. We computed β for viable monochorionic (MC) and dichorionic (DC) twin pregnancies [35]. In DC twins, each sibling has its "own" placenta, while MC twins share a single placenta. β was found to be 0.76 ± 0.02 for both MC and DC twins. This range is very close to what was found for term singleton pregnancies. Finally, while the wild-type mouse placenta is structurally very different from the human, the same scaling exponent relates mouse fetal and placental weights but only after the time when the branched labyrinth has developed [36].

The consistency of the scaling factor values in relatively well populations of singletons, twins, and mice makes β, the ubiquitous exponent of Kleiber's law, a useful proxy for understanding pathologic placental structure, and by extension, function in singleton and twin human pregnancies.

2.2 Chorionic plate vasculature

From the umbilical cord insertion on the chorionic plate extends a highly variable network of arterial and venous high capacitance/low resistance vessels. The chorionic surface arteries branch repeatedly, finally exiting the chorionic plate from the network of fetal stem vessels and terminal villi filled with capillaries that carry fetal blood into proximity to the maternal blood in the intervillous space. The capillaries then unite to form venules and join the chorionic surface veins that carry the oxygenated blood and nutrients to the fetus. Unlike other regions in the human body, the chorionic arterial and venous networks do not run in parallel. The basis(es) of these variations is unknown but is hypothesized to be local adaptations to intrauterine variations in the decidua, vasculature, and/or local inflammation.

A search for fractal relationships in the whole placenta has been done using MRI and X-ray angiograms of casted placenta fact [37, 38]. However, using MRI and casting the placenta for X-ray angiograms are not trivial and cannot be routinely performed. Therefore, we elected to study placental fractals with hand-traced placental chorionic surface vessel networks from digital 2D images of post-delivery placentas (**Figure 2**) [39]. From such images of chorionic vessels, their length and

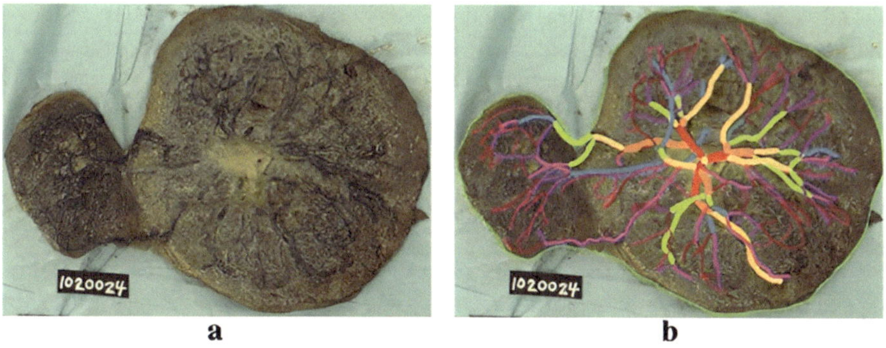

Figure 2.
a) 2D image of the fetal surface of placenta, b) placental surface vascular tracing based on protocol described in [39]. Source: Adapted from [27].

branching characteristics—the number of branch points, branching diameters, vessel paths, branching angles, tortuosity, and branching generations—can be extracted [27]. The mean fractal dimension of the chorionic surface arterial network computed from binary images of the tracings of normal term pregnancies was similar to that of term pregnancies with diabetes and preeclampsia but differed from that of preterm pregnancies [12]. The number of arterial branch points per unit of chorionic plate area of placentas of uncomplicated pregnancies differed from those of pregnancies with diabetes or preeclampsia [12]. The fact that the fractal dimension of normal pregnancies differed from that of preterm pregnancies raises the question of whether preterm birth could be a consequence of sub-optimal placental function.

Another study compared the fractal dimension of placental chorionic surface arterial networks from 197 normal singleton pregnancies and 60 monochorionic twins (twins sharing the same placenta). The twins' group was subdivided into two groups: one with significant BW discordance (BW of one twin is greater than the other one by >20%) and the other without BW discordance. BW discordance in twins has been associated with neonatal mortality risk. It was found that the fractal dimension of singleton pregnancies was significantly lower from twins, both with and without BW discordance [40]. On the other hand, the branch points per unit area between the term singleton and the twin placentas remained the same. These findings suggest that despite having different intrinsic branching characteristics of placental chorionic surface vasculature, the singleton and twin pregnancies may have a similar "angiogenic potential," which can regulate and/or limit their overall growth.

Placental chorionic surface vessel features have also shown differences between the term "well" general population pregnancies and high autism-risk pregnancies (subsequent pregnancies of mothers that already have children with autism diagnosis). The chorionic surface vessels of 89 pregnancies from the high autism-risk Early Autism Risk Longitudinal Investigation (EARLI) cohort, in general, had fewer branch points, thicker and less tortuous arteries, better extension to the surface boundary, and smaller branch angles as compared to 201 pregnancies from the general population-based National Children's Study.

Another approach to evaluate the spatial distribution and development across the gestation of placental surface vasculature is to employ Voronoi diagrams. A Voronoi diagram is a partition of a plane into regions close to each of a given set of objects [41]. In the simplest case, these objects are just finitely multiple points in the plane (called seeds, sites, or generators). For each seed, there is a corresponding region, called a Voronoi cell, consisting of all points of the plane closer to that seed than to any other. Voronoi diagrams have practical and theoretical applications in various fields, primarily science and technology [42, 43]. We compared the surface vascular spatial distribution of the high autism-risk EARLI cohort and the general population-based National Children's Study cohort by dividing the chorionic plate area into Voronoi cells. The Voronoi cells were computed by choosing the midpoints of placental surface vascular branches as Voronoi seeds. The histograms of Voronoi cell areas were plotted, and it was found that the scaled distributions of Voronoi cell areas in the two cohorts collapsed into a single distribution, but the EARLI cohort showed a lower branching density [44]. The fact that both the cohorts showed similar spatial distribution of the surface vessels could indicate that the overall mechanism for the formation and development of the vasculature is the same in normal and high autism-risk pregnancies, but is less active in the high autism-risk pregnancies, yielding a lower branching density.

The fractal nature of the chorionic plate vessels is thought to be important for multiple reasons. Not only does it help to increase the surface area of the vessel, allowing for a more efficient gas exchange between the fetus and the mother, but it also helps distribute the blood optimally and protects the vessels from injury. Deviations of placental vasculature from their standard fractal nature can help to identify potential pathological outcomes and also understand the underlying mechanisms across the gestation that potentially influence the child health outcome.

2.3 Placental shape: 2D fractal models

An average delivered human placenta is more or less round, with a centrally inserted umbilical cord and has a relatively uniform thickness. However, a fair number of placentas are irregularly shaped [45] and can be classified into several well-defined geometrical patterns (**Figure 3**). This is not true for our "permanent" organs (heart, lungs, etc.), which have standard shapes, and any deviations result in

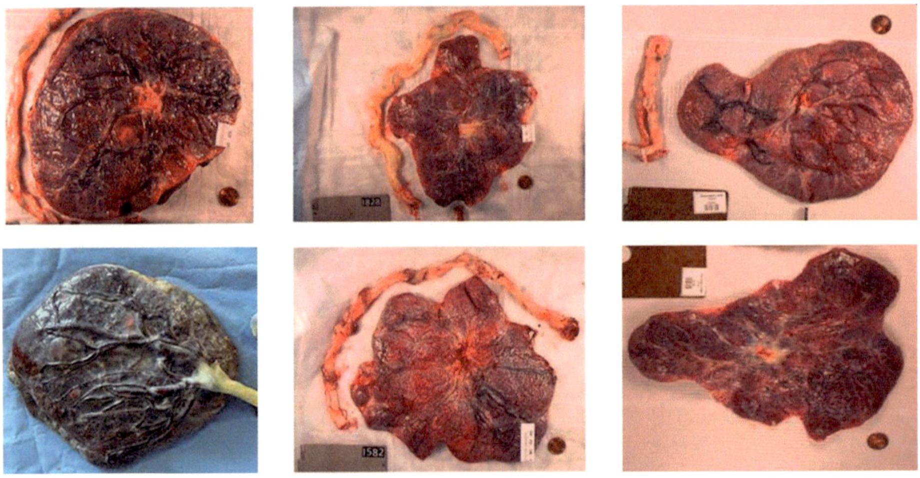

Figure 3.
Multiple shapes of the human placenta. The left column shows round/oval-shaped placentas, the middle column shows star-shaped placentas, and the right column shows multilobed placentas. Adapted from [46].

obvious dysfunction. In the National Collaborative Perinatal Project data, irregular placental shapes were associated with lower BW for placental weight, suggesting variably shaped placentas have altered function. The traditional measurements of placental shape in terms of major and minor axes and average placental thickness are highly subjective. Technological advances have made it exceedingly easy to capture a simple photograph of the placenta with a unit distance marker such as a ruler (**Figure 3**). Such photographs have allowed precise measurements of placental shape (deviation from roundness), cord eccentricity (how central or noncentral the umbilical cord insertion is), area, perimeter, and so on. [16]. Placental efficiency has been suggested to be affected by patterns of placental growth, specifically, the relationship between chorionic plate area and disk thickness [16, 34, 46, 47].

Empirical models based on diffusion-limited aggregation have suggested specific shape variations to have their origins at distinct periods in gestation (**Figure 4**) [46, 49]. These models have also accounted for multilobed and star-shaped placentas (two of the most common patterns of abnormally shaped placentas). Also, they have provided empirical evidence that the shape of the placenta does reflect the underlying vascular fractal; deviations from the umbilical cord insertion are associated with reduced placental functional efficiency. These deviations are contemplated to reflect placental compensations responding to stressors or variability in the intrauterine environment. Studies have also shown that chorionic plate measures by ultrasound at 11–14 weeks are significantly correlated with similar features observed at term [49]. Thus, these shape measures can provide insights into the impacts of intrauterine environmental influences at different gestational points that lead up to the "final" placenta at delivery and help study their temporal and collective impacts on the child's health.

Digital photographs of the fetal placenta surface and the sliced placental disk from 129 high autism-risk newborns in the EARLI cohort and 267 newborns in the National Children's Study Vanguard pilot were analyzed to extract comparable measures of placental chorionic surface shape, umbilical cord displacement, and disk thickness

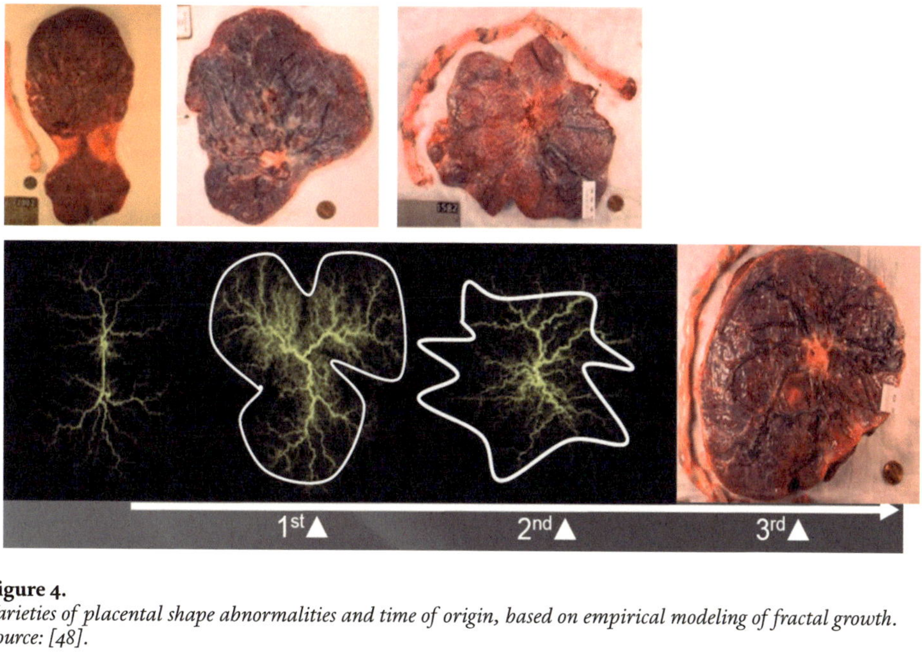

Figure 4.
Varieties of placental shape abnormalities and time of origin, based on empirical modeling of fractal growth. Source: [48].

[50]. Placental thickness measures were moderately higher in high autism-risk cases. The placentas of high autism-risk pregnancies were also rounder and more regular in perimeter than general population placentas. This data were in line with the evidence presented from a small nested case-control comparison of 52 autism cases and 161 controls from the Avon Longitudinal Study of Parents and Children general population pregnancy cohort study, which showed that children who would receive an autism diagnosis had reduced cord insertion eccentricity and reduced villous branching growth as compared to their sex and gestational age-matched peers [51]. Reduced placental shape variability observed in high ASD-risk siblings compared to low-risk controls may indicate a restricted ability to compensate for intrauterine changes. In other words, the placentas of children with a high autism risk prefer a uniform intrauterine environment and do not respond effectively to any stressors, which could lead to a sub-optimal placenta and, by extension, a poor neurodevelopmental outcome.

2.4 Placental shape: 3D

While placental shape defined by the chorionic plate aspect ratio and placental thickness from slices of placenta have provided invaluable insights into the relations between placental shape and function, the placenta is a three-dimensional organ, and a 3D digital image may be more useful to understand the timing and cause of variations in placental shape. The 3D image can provide the overall placental volume and also help in visualizing and analyzing the spatial distribution of placental thickness—the primary dimension of placental growth in the third trimester [52]. Additionally, we can proxy time order in the placenta by assigning x, and y coordinates referencing the umbilical cord as the site of origin (0,0), with time order inferred from increasing distance from the umbilical cord (**Figure 5**).

Our group used a 3D surface scanner to scan 264 placentas from the National Children's Study and 105 placentas from the high-autism EARLI cohort [53]. They computed placental volume and the maximum radius from the umbilical insertion point and volumes encompassed within each of the ten concentric circles with radii at each decile % of the maximum radius (**Figure 6**). Two slopes were then computed for placental volume distribution during two time periods: early placental growth (1–6 deciles) and late growth (7–10 deciles). The placental volume and its growth trajectory differed between normal and high autism-risk pregnancies. Early growth trajectory

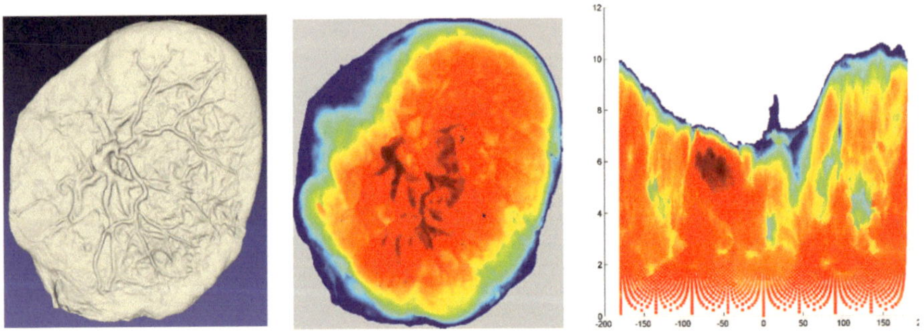

Figure 5.
Fetal side of a placental 3D scan (left); thickness map of the placental 3D scan with thicker portions shown in red and thinner portions shown in green and blue (middle); polar map conversion of thickness map where the insertion point is along the x-axis and the y-axis demonstrates the radial growth in different directions over gestation (right).

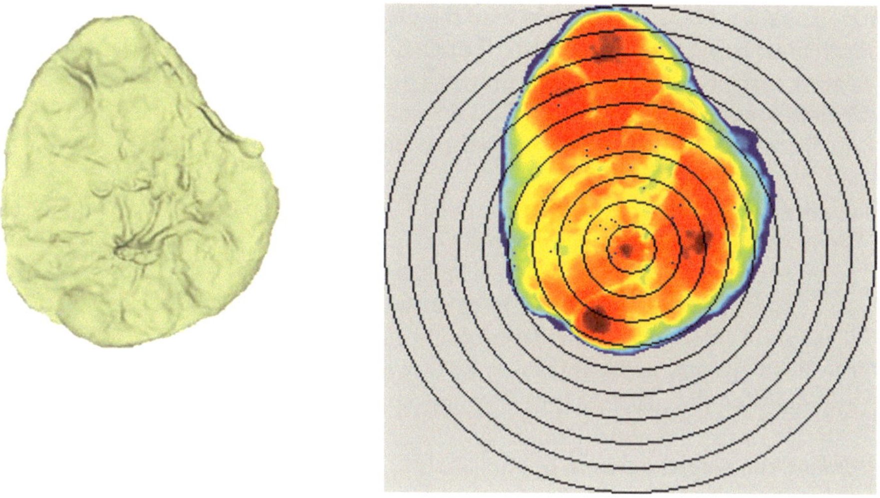

Figure 6.
Fetal side of a placental 3D scan (left); deciles drawn over thickness map of the placenta based on the maximum radius to determine volume growth over the gestation (right). Source: [53].

of high autism-risk placentas showed sex-dimorphism, indicating an underlying mechanism of placental growth that differentiates boys and girls, which may lead to a differential rate of autism diagnosis between them.

2.4.1 Spatial analysis of the placenta using geographical information system (GIS) theory

What the stand-alone 2D and 3D approaches lack is a true cohesive method to quantify the observed patterns/deviations in the placental structure at different scales and, by extension, explore the processes that may have generated the patterns/deviations. Spatial analysis helps define attributes in the context of their location and neighborhood and helps explore spatial patterns and relationships of data points, features, or regions in a given area. Geographical information system (GIS) provides the tools to capture, store, manipulate, analyze, and visualize spatial data. Apart from geographical applications such as studying land cover patterns and river trajectories [54, 55], spatial analysis has found biological applications such as identifying spatial patterns of neuronal activity, muscle fibers, and gene expression [56–58]. The fractal nature of GIS arises from the fact that spatial phenomena often exhibit self-similarity across multiple scales, and GIS is designed to represent and analyze such phenomena at different scales of analysis. One way to understand the fractal nature of GIS is to consider how it can be used to analyze land cover patterns. Land cover patterns, such as forests, grasslands, and urban areas, often exhibit fractal structures, with similar patterns appearing at different scales of analysis. GIS can be used to capture and analyze these patterns at different scales, from the individual tree or building to the landscape or region as a whole.

GIS theory can be applied to study the placental topography from the placental 2D images and 3D scans. The placental topography can be compared to a country or state's landscape, while the placental vasculature compares to river networks. In typical GIS mapping, exposures such as air pollution are assigned to geographic sites (a county or region); site, exposure, and even change over time can be considered jointly in analysis. We can similarly map the placental chorionic plate as sets of x and y coordinates, with assigned local attributes such as disk thickness, and distance to the nearest surface vessel or branch points. This approach promises to greatly improve our ability to model the placenta in fetal origins research.

Implementing GIS broadly involves three steps:

i. **define the spatial entity**: We first need to define the spatial unit on which the analysis is to be conducted. The placental topography can be divided into units (akin to counties on a state map). The placental 3D image can be divided based on a raster (i.e., divide a map into equal squares and check adjoining squares, or it can be divided in a biological context e.g., registering the 2D surface vessels on the 3D image and using chorionic surface arterial branch points as Voronoi seeds and dividing the placenta into Voronoi cells) (**Figure 7**)

ii. **define the neighborhood**: Spatial analysis helps to explore patterns and relationships of attributes in the context of its location and neighborhood; for example, how does mean income compare in the tri-state area? Once the spatial entity is established, it is important to figure out the neighbors with whom the attribute value is to be compared (**Figure 8**). One option is to consider all the neighbors with adjoining Voronoi cell borders. Another option is to select all Voronoi cells

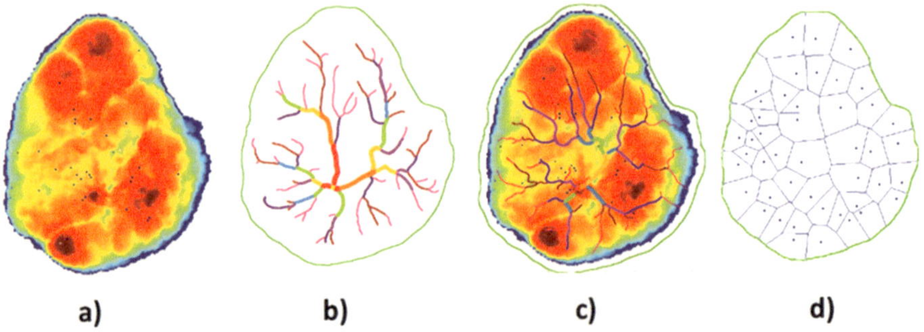

Figure 7.
Dividing placental topography. a) 3D placental thickness map, b) 2D surface vessel tracing, c) registered output of 2D vessels overlaid on the 3D thickness map, d) division of placental surface into Voronoi cells by using the branch points as Voronoi seeds.

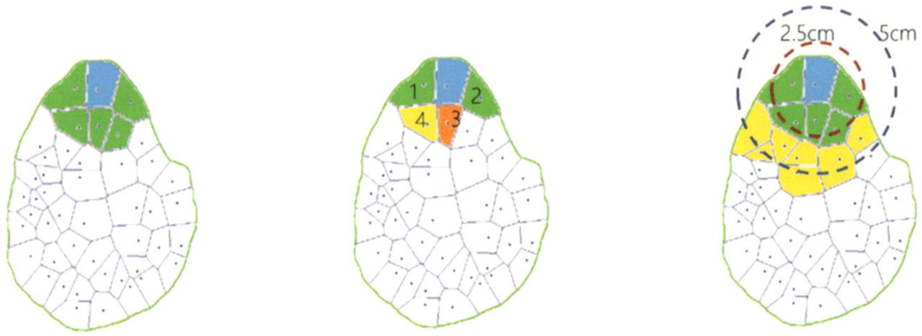

Figure 8.
Neighborhood selection for GIS-mapping of the placenta (blue cell is the Voronoi cell of interest). All adjacent neighbors that share a boundary, as shown by green cells (left), k-nearest neighbors (middle), neighbors determined by radial distance (right).

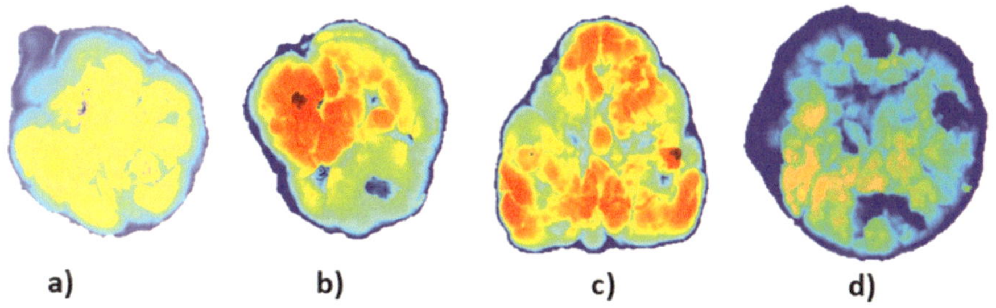

Figure 9.
Different patterns of placental thickness distribution. a) Uniform thickness distribution will have high Moran's I for near and distant neighbors, b) clustered thickness distribution will have high Moran's I for near neighbors and lower Moran's I for distant neighbors, c) patches or starfish-like pattern will have lower Moran's I for near neighbors and higher Moran's I for distant neighbors, d) random thickness distribution will have lower Moran's I for both near and distant neighbors.

within a particular radial band (e.g., 2 cm, 5 cm). A third option is to select a fixed neighbor of nearest neighbors (e.g., 2, 3 or 4 nearest neighbors).

iii. **using a spatial analysis statistic to compute the spatial distribution**: Moran's I is the most commonly used spatial autocorrelation measure [59]. Like other correlation coefficients, Moran's I ranges between −1 and 1. Autocorrelation is high when neighboring physical sites share similar attributes and low when there is variability in those attributes. Moran's I can therefore provide indices of placental structural simplicity or uniformity (reflected in positive Moran's I and high autocorrelation) or their structural complexity or variability (represented variably by patterns of positive or negative Moran's I, in either or both global and local autocorrelation) (**Figure 9**). Thus, a range of Moran's I capture complex aspects of placental topography and thereby provide an important window into understanding the intrauterine environment across gestation.

Since the placenta grows outward from the umbilical insertion point across the gestation, GIS also allows for analysis of rates of change of placental features over time. Combining loci with attributes related to cord insertion, the placenta can be modeled in its full spatial and temporal context. Under the NIH's Environmental influences on Child Health Outcomes (ECHO) program, we are currently developing GIS-based models of the placenta to study if prenatal stressors (maternal depression and stress) impact child well-being and neurodevelopment via the mediating effects of the placenta.

2.5 Villous spatial distribution

Terminal villi are the tips of the chorionic villi that have matured to have a thin trophoblast epithelium, which is in direct contact with the maternal intervillous blood and cores of villous stroma, including capillaries, fibroblasts, and Hofbauer cells, the placenta's resident immune cell. These terminal villi are the surface area for the exchange of nutrients, oxygen, and waste products between the mother and fetus.

The marked variation in the appearance of the villi has been well appreciated irrespective of the placental size. The variation in villous distribution (measured as both packing density and lacunarity) has been demonstrated to impact the placental efficiency [60]. This histologic variation is likely associated with the high capacity of the placenta for genetic variation, in terms of chromosomal mosaicism and epigenetic variation (in terms of DNA methylation). PW depends on villous histology, including both villous structure and spatial distribution. Morphometric assessment of placental mature villi in healthy, smoking and non-smoking women demonstrated that the fractal dimension increased with the number of cigarettes smoked per day [28]. Another study computed the fractal dimension of villi from placental histology slides of different grades of distal villous hypoplasia (a correlate of fetal growth restriction) and compared them to the average grading of expert pathologists. The fractal dimension decreased as the grade of distal villus hypoplasia increased and showed a fair correlation with the experts' grades [29]. Thus, the fractal dimension helps to capture the variation in the structural and spatial distribution of the placental villi that may be a result of exposures such as smoking during pregnancy or due to an underlying pathological condition.

The fractal nature of the placental villi is thought to be an adaptation that maximizes the exchange of materials between the mother and fetus, allowing for an efficient transfer of nutrients and oxygen to support fetal growth and development. A quantitative assessment of the spatial distribution of placental villi and their fractal properties can help distinguish sub-optimal systems that lead to poor pregnancy outcomes.

3. Discussion

The placenta has been shown to have a fractal nature at different scales. This has been observed in the allometric scaling of its weight and infant BW, its macroscopic shape and surface vasculature, and its microscopic villi shape and distribution. Fractal analysis of the placenta at these various scales can provide an improved understanding of the pregnancy and also help early identification of potential risk factors.

The allometric scaling relationship between the fetal and placental weights can help identify sub-optimal placentas. The fact that the allometric scaling holds true for both singleton and twin pregnancies demonstrates that all pregnancies could be governed by the same underlying capacity. The cord insertion and underlying vasculature influence the placental growth early in gestation and can determine the overall placental shape at delivery. The rate of placental growth and function varies across gestation and can be captured with the 3D placental volume trajectories going out from the umbilical cord insertion. The fractal nature of the plate and its vasculature is highly influenced by the intrauterine exposures and gives a glimpse into the placental growth and its effort in trying to reach the optimal morphology and function across the gestation. Deviations of this fractality could be used as early markers of potentially risky pregnancy outcomes. The sexual dimorphism of neurodevelopmental outcomes such as autism is also captured by the placental shape and vasculature fractality and could be used as early biomarkers to identify the risk of such outcomes and develop early intervention plans to help the child and family. The microscopic villi, which are the functional units of the placenta, are routinely captured by pathologists on a regular histology slide. The fractal dimensions of these villi have been observed to vary with poor pregnancy outcomes and could be an invaluable quantification tool to complement the histopathological evaluation of pathologists. Studying the spatial distribution and the growth trajectory of the placenta by combining the chorionic plate vasculature data with the 3D shape of the placenta can provide a great model for examining the placental growth in the context of the intrauterine environment across gestation.

Thus, investigating the fractal nature of the placenta at different scales can illuminate the mechanisms of its growth and function and also helps to identify early risks of poor pregnancy outcomes.

4. Conclusion

The placenta is a complex and highly organized structure that facilitates the exchange of nutrients and waste between the mother and fetus. By studying its fractal nature, we can gain a better understanding of its branching patterns and organization at different scales. This knowledge can help us identify potential abnormalities and develop new treatments for placental disorders. Research has shown that the fractal dimension of the placenta is related to fetal growth and BW. By analyzing the fractal properties of the placenta early in pregnancy, we may be able to predict the risk of adverse pregnancy outcomes, such as preterm birth or fetal growth restriction. Fractal analysis can be used to develop new imaging techniques that can provide detailed information about placental structure and function. For example, researchers can use fractal analysis to create three-dimensional models of the placenta that can be used for virtual reality simulations and surgical planning. The placenta is quintessential for fetal development. By understanding the fractal nature of the placenta, we may be able to gain a better understanding of how the placenta develops and functions over time, which could lead to new insights into fetal development.

Overall, studying the fractal nature of the placenta can provide valuable insights into its structure, function, and development and may ultimately lead to new treatments for placental disorders and better pregnancy outcomes.

Acknowledgements

The GIS-based method developed to study the placental topography that is reported in this chapter was supported by the Environmental influences on Child Health Outcomes (ECHO) Program, Opportunity and Infrastructure Fund Award Number U2COD023375.

Conflict of interest

The authors declare no conflict of interest.

Author details

Ruchit Shah[1,2], Carolyn Salafia[1,2*], Theresa Girardi[1,2], Pratipaul Persaud[1,2], Jessica Brunner[3], Emily S. Barrett[4], Thomas O'Connor[5] and Richard K. Miller[3]

1 Placental Analytics, LLC, New Rochelle, USA

2 New York State Institute for Basic Research, Staten Island, USA

3 Department of Obstetrics and Gynecology, School of Medicine and Dentistry, University of Rochester, Rochester, USA

4 Department of Biostatistics and Epidemiology, Rutgers School of Public Health, Environmental and Occupational Health Sciences Institute, Piscataway, USA

5 Department of Psychiatry, School of Medicine and Dentistry, University of Rochester, Rochester, USA

*Address all correspondence to: carolyn.salafia@gmail.com

References

[1] Glenny RW, Robertson HT, Yamashiro S, Bassingthwaighte JB. Applications of fractal analysis to physiology. Journal Of Applied Physiology Bethesda Md 1985. 1991;**70**(6):2351-2367

[2] El-Nabulsi RA, Anukool W. Fractal dimensions in fluid dynamics and their effects on the Rayleigh problem, the Burger's vortex and the kelvin–Helmholtz instability. Acta Mech. 2022;**233**(1):363-381

[3] Pfeifer P, Avnir D. Chemistry in noninteger dimensions between two and three. I. Fractal theory of heterogeneous surfaces. The Journal of Chemical Physics. 1983;**79**(7):3558-3565

[4] Mandelbrot BB. The Fractal Geometry of Nature. New York: Henry Holt and Company; 1983. p. 504

[5] Sen J, McGill D. Fractal analysis of heart rate variability as a predictor of mortality: A systematic review and meta-analysis. Chaos Woodbury N. 2018;**28**(7):072101

[6] Perkiömäki JS, Mäkikallio TH, Huikuri HV. Fractal and complexity measures of heart rate variability. Clinical and Experimental Hypertension N Y N 1993. 2005;**27**(2-3):149-158

[7] Bayat S, Broche L, Dégrugilliers L, Porra L, Paiva M, Verbanck S. Fractal analysis reveals functional unit of ventilation in the lung. The Journal of Physiology. 2021;**599**(22):5121-5132

[8] Yeh JR, Sun WZ, Shieh JS, Huang NE. Investigating fractal property and respiratory modulation of human heartbeat time series using empirical mode decomposition. Medical Engineering & Physics. 2010;**32**(5):490-496

[9] Delides A, Viskos A. Fractal quantitative endoscopic evaluation of the upper airway in patients with obstructive sleep apnea syndrome. Otolaryngology--Head Neck Surgery Official Journal of American Academic Otolaryngology-Head Neck Surgery. 2010;**143**(1):85-89

[10] Lemmens S, Luyts M, Gerrits N, Ivanova A, Landtmeeters C, Peeters R, et al. Age-related changes in the fractal dimension of the retinal microvasculature, effects of cardiovascular risk factors and smoking behaviour. Acta Ophthalmologica. 2022;**100**(5):e1112-e1119

[11] Masters BR. Fractal analysis of the vascular tree in the human retina. Annual Review of Biomedical Engineering. 2004;**6**:427-452

[12] Shah R, Girardi T, Ma X, Salafia C. Fractal dimensions and branching characteristics of placental chorionic surface arteries. Placenta. 2018;**70**:4-6

[13] Smith JH, Rowland C, Harland B, Moslehi S, Montgomery RD, Schobert K, et al. How neurons exploit fractal geometry to optimize their network connectivity. Scientific Reports. 2021;**11**(1):2332

[14] Nguyen TT, Hoffmann E, Buerkert A. Spatial patterns of urbanising landscapes in the north Indian Punjab show features predicted by fractal theory. Scientific Reports. 2022;**12**(1):1819

[15] Fernández E, Bolea JA, Ortega G, Louis E. Are neurons multifractals?

Journal of Neuroscience Methods. 1999;**89**(2):151-157

[16] Salafia CM, Yampolsky M, Misra DP, Shlakhter O, Haas D, Eucker B, et al. Placental surface shape, function, and effects of maternal and fetal vascular pathology. Placenta. 2010;**31**(11):958-962

[17] Salafia CM, Thomas DM, Roberts DJ, Straughen JK, Catalano PM, Perez-Avilan G. First trimester detection of placental disease: Challenges and opportunities. American Journal of Perinatology. 2016;**33**(13):1306-1312

[18] Longtine MS, Nelson DM. Placental dysfunction and fetal programming: The importance of placental size, shape, histopathology, and molecular composition. Seminars in Reproductive Medicine. 2011;**29**(3):187-196

[19] Salafia CM, Shah RG, Misra DP, Straughen JK, Roberts DJ, Troxler L, et al. Chorionic vascular "fit" in the human placenta: Relationship to fetoplacental outcomes. Placenta. 2017;**59**:13-18

[20] Dahdouh S, Andescavage N, Yewale S, Yarish A, Lanham D, Bulas D, et al. In vivo placental MRI shape and textural features predict fetal growth restriction and postnatal outcome. Journal of Magnetic Resonance Imaging JMRI. 2018;**47**(2):449-458

[21] Salavati N, Sovio U, Mayo RP, Charnock-Jones DS, Smith GCS. The relationship between human placental morphometry and ultrasonic measurements of utero-placental blood flow and fetal growth. Placenta. 2016;**38**:41-48

[22] Cotechini T, Hopman WJ, Graham CH. Inflammation-induced fetal growth restriction in rats is associated with altered placental morphometrics. Placenta. 2014;**35**(8):575-581

[23] Cotter SL, Klika V, Kimpton L, Collins S, Heazell AEP. A stochastic model for early placental development. Journal of the Royal Society Interface. 2014;**11**(97):20140149

[24] Palomba S, Russo T, Falbo A, Di Cello A, Tolino A, Tucci L, et al. Macroscopic and microscopic findings of the placenta in women with polycystic ovary syndrome. Human in Reproduction Oxford England. 2013;**28**(10):2838-2847

[25] Winder NR, Krishnaveni GV, Veena SR, Hill JC, Karat CLS, Thornburg KL, et al. Mother's lifetime nutrition and the size, shape and efficiency of the placenta. Placenta. 2011;**32**(11):806-810

[26] Yampolsky M, Salafia CM, Misra DP, Shlakhter O, Gill JS. Is the placental disk really an ellipse? Placenta. 2013;**34**(4):391-393

[27] Chang JM, Zeng H, Han R, Chang YM, Shah R, Salafia CM, et al. Autism risk classification using placental chorionic surface vascular network features. BMC Medical Informatics and Decision Making. 2017;**17**(1):162

[28] Wielgus E, Pawlicki K, Kawa A, Włoch S, Kamiński M. Fractal analysis of placenta mature villi in healthy, smoking and non-smoking women. Medical Science Monitor: International Medical Journal of Experimental and Clinical Research. 2000;**6**(2):271-277

[29] Mukherjee A, Chan ADC, Keating S, Redline RW, Fritsch MK, Machin GA, et al. The placental distal villous hypoplasia pattern: Interobserver agreement and automated fractal dimension as an objective metric. Pediatric and Developmental Pathology Official Journal of Society Pediatric

Pathology Paediatric Pathology Society. 2016;**19**(1):31-36

[30] Kleiber M. The fire of life: An introduction to animal energetics. New York: Wiley; 1961

[31] West GB, Brown JH, Enquist BJ. A general model for the origin of allometric scaling laws in biology. Science. 1997;**276**(5309):122-126

[32] Brown E by JH, West GB, editors. Scaling in Biology. Oxford, New York: Oxford University Press; 2000. 368 p. (Santa Fe Institute Studies on the Sciences of Complexity).

[33] Gruenwald P. The Placenta and its Maternal Supply Line: Effects of Insufficiency on the Fetus. Baltimore: University Park Press; 1975

[34] Salafia CM, Misra DP, Yampolsky M, Charles AK, Miller RK. Allometric metabolic scaling and fetal and placental weight. Placenta. 2009;**30**(4):355-360

[35] Salafia CM, Kiryankova N, Inany H, Charlagorla P, Park M, Khawar N, et al. Metabolic scaling and twin placentas. Placenta. 2016;**37**:16-18

[36] Gasperowicz M, Yampolsky M, Salafia CM. Metabolic scaling law for mouse fetal and placental weight. Placenta. 2013;**34**(11):1099-1101

[37] Bergman DL, Ullberg U. Scaling properties of the placenta's arterial tree. Journal of Theoretical Biology. 1998;**193**(4):731-738

[38] Parker KJ, Carroll-Nellenback JJ, Wood RW. The 3D spatial autocorrelation of the branching fractal vasculature. Acoust Basel Switz. 2019;**1**(2):369-382

[39] Shah RG, Salafia CM, Girardi T, Conrad L, Keaty K, Bartleotc A. Shape matching algorithm to validate the tracing protocol of placental chorionic surface vessel networks. Placenta. 2015;**36**(8):944-946

[40] Shah RG, Girardi T, Salafia CM, Lopriore E, Zhou D, Nakamura M. Fractal dimensions of placental chorionic surface arteries of monochorionic twins. Reproductive Sciences. 2019;**26**(1)(Suppl. 1):170A-170A

[41] Voronoi G. Nouvelles applications des paramètres continus à la théorie des formes quadratiques. Deuxième mémoire. Recherches sur les parallélloèdres primitifs. Journal für die reine und angewandte Mathematik (Crelles Journal). 1908;**134**:198-287

[42] Sen Z. Spatial Modeling Principles in Earth Sciences. Switzerland: Springer; 2016. p. 424

[43] Okabe A, Boots B, Sugihara K, Chiu SN. Spatial Tessellations: Concepts and Applications of Voronoi Diagrams. 2nd ed. Chichester, England: Wiley; 2000

[44] Leonard AS, Lee J, Schubert D, Croen LA, Fallin MD, Newschaffer CJ, et al. Scaling of the surface vasculature on the human placenta. Physical Review E. 2017;**96**(4-1):040401

[45] Baergen RN. Macroscopic evaluation of the second- and third-trimester placenta. In: Baergen RN, editor. Manual of Pathology of the Human Placenta: Second Edition [Internet]. Boston, MA: Springer US; 2011. pp. 23-42. [cited 2023 May 7] Available from:. DOI: 10.1007/978-1-4419-7494-5_3

[46] Yampolsky M, Salafia CM, Shlakhter O, Haas D, Eucker B, Thorp J. Modeling the variability of shapes of a human placenta. Placenta. 2008;**29**(9):790-797

[47] Salafia CM, Yampolsky M. Metabolic scaling law for fetus and placenta. Placenta. 2009;**30**(5):468-471

[48] Salafia C, Roberts DJ. Placental weight, shape and gross vascular morphology. In: Khong TY, Mooney EE, Nikkels PGJ, Morgan TK, Gordijn SJ, editor. Pathology of the Placenta: A Practical Guide. Cham: Springer International Publishing; 2019. pp. 47-54. DOI: 10.1007/978-3-319-97214-5_5

[49] Salafia CM, Yampolsky M, Shlakhter A, Mandel DH, Schwartz N. Variety in placental shape: When does it originate? Placenta. 2012;**33**(3):164-170

[50] Park BY, Misra DP, Moye J, Miller RK, Croen L, Fallin MD, et al. Placental gross shape differences in a high autism risk cohort and the general population. PLoS One. 2018;**13**(8):e0191276

[51] Salafia C, Misra D, Golding J, Platt C, Ring S. Characterization of placental growth as a biomarker of autism/ASD risk. Placenta. 1 Sep 2012;**33**(9):A16

[52] Moore KL, Persaud TVN, Torchia MG. The Developing Human - 9780323611541 [Internet]. 10th ed. Philadelphia: Saunders; 2015 [cited 2023 May 7]. Available from: https://www.us.elsevierhealth.com/the-developing-human-9780323611541.html

[53] Shah R, Salafia C, Girardi T, Miller R, Moye J, Lyall K, et al. Difference in placental growth volume in Normal and high autism-risk pregnancies. In: Reproductive Sciences. 2455 Teller RD, Thousand Oaks, CA 91320 USA: Sage Publications Inc; 2019. p. 174A 174A

[54] Cheng L, Feng R, Wang L. Fractal characteristic analysis of urban land-cover spatial patterns with spatiotemporal remote sensing images in Shenzhen City (1988-2015). Remote Sensing. 2021;**13**(22):4640

[55] Rodríguez-Iturbe I, Rinaldo A. Fractal River Basins: Chance and Self-Organization [Internet]. [cited 2023 May 7]. Available from: https://www.amazon.com/Fractal-River-Basins-Chance-Self-Organization/dp/0521004055

[56] Achim K, Pettit JB, Saraiva LR, Gavriouchkina D, Larsson T, Arendt D, et al. High-throughput spatial mapping of single-cell RNA-seq data to tissue of origin. Nature Biotechnology. 2015;**33**(5):503-509

[57] Millet LJ, Collens MB, Perry GLW, Bashir R. Pattern analysis and spatial distribution of neurons in culture. Integrative Biology: Quantitative Biosciences From Nano to Macro. 2011;**3**(12):1167-1178

[58] Bodine-Fowler S, Garfinkel A, Roy RR, Edgerton VR. Spatial distribution of muscle fibers within the territory of a motor unit. Muscle & Nerve. 1990;**13**(12):1133-1145

[59] Moran P, a. P. Notes on continuous stochastic phenomena. Biometrika. 1950;**37**(1-2):17-23

[60] Shah R, Salafia C, Girardi T, Merz G. Villus packing density and lacunarity: Markers of placental efficiency? Placenta. 2016;**48**:68-71

Chapter 9

Multiscaling Description of the Space-Time Structure of Rainfall

Victor M. Peñaranda-Vélez, Carlos E. Puente, Óscar J. Mesa and Arturo I. Quintanar

Abstract

Understanding the space-time structure of rainfall is one of the most complex problems faced in geophysics and atmospheric sciences. Rainfall has been studied in a similar fashion than turbulence phenomena and many of its mathematical tools have been applied to the description of rainfall patterns. Some of the tools used in turbulence phenomena and rainfall were born in the theory of fractal geometry. Such a theory has turned out to be the most suitable scientific toolbox for explaining complexity, nonlinear processes, and multiscaling dynamics. The aim of this chapter is to illustrate some of these ideas related to the statistical characterization of the space-time structure of rainfall, including the proposals for modeling rainfall fields and some achievements in the construction of a holistic multiscaling theory for rainfall field description.

Keywords: multifractals, multiscaling, multiplicative processes, fractal-multifractal method, space-time, rainfall

1. Introduction

Rainfall can be described as irregular patterns that look random, and such a behavior has been mostly associated with the one found in turbulence phenomena. The statistical structure of rainfall is enriched by many self-similar fluctuations of different sizes, which possess a particular fractal dimension; besides, its structure is also characterized by a property known as *scaling invariance* that explains how rainfall patterns can be understood as *self-similar* objects. After many years of studying rainfall, the *multifractal* approach has been the most suitable tool for understanding how rainfall behaves in a statistical sense and also in a physical sense. From this approach, *multifractal measures* are widely used to study highly irregular distributions of physical quantities defined on a geometric support [1, 2].

Originally, the application of the multifractal approach for studying physical systems was born in the study of *turbulence phenomena* [3–6]. The techniques and mathematical tools used in such phenomena were subsequently extended to the study of rainfall. On the study of fully developed turbulent flows, the starting point is in the Navier-Stokes equations:

$$\partial_t \mathbf{u} + \mathbf{u}\partial_x \mathbf{u} = \partial_x p + \nu \partial_{xx}\mathbf{u}; \quad \partial_x \mathbf{u} = 0; \quad \mathbf{u}(0, \mathbf{x}) = u_0 \tag{1}$$

in which a broken symmetry property has been identified on them. One can identify on these equations that their scaling properties are represented by infinitely many scaling groups, particularly when the viscous term vanishes. Moreover, the Navier-Stokes equations are invariant under scaling transformations g_λ, that is:

$$g_\lambda : \quad \mathbf{x} \to \lambda \mathbf{x}; \quad \mathbf{u} \to \lambda^H \mathbf{u}; \quad t \to \lambda^{1-H} t \tag{2}$$

where H represents a *self-similarity scaling exponent*. As the energy production is given at a unique scale λ_0 (integral scale), the invariance property cannot be sustained for all scales. Kolmogorov's theory [7] postulates that there is a finite rate of energy dissipation per unit mass in the turbulent flow, and that such a flow is statistically self-similar with a scaling exponent $H = 1/3$.

Previous studies about intermittency in fully developed turbulent flows use the *structure function* to identify the self-similarity scaling exponent. For instance, the (longitudinal) structure function $S_q(\ell)$ is defined as the q-th order moment of the velocity fluctuation on a distance ℓ, this is:

$$S_q(\ell) \equiv \langle |\mathbf{u}(\mathbf{x} + \ell) - \mathbf{u}(\mathbf{x})|^q \rangle \equiv \langle |\Delta \mathbf{u}(\ell)|^q \rangle \tag{3}$$

where q is the order of the moment, $\mathbf{u}(\mathbf{x})$ is the velocity at location $\mathbf{x}$, and the operator $\langle \cdot \rangle$ represents an average. For homogeneous isotropic turbulence, the structure function follows a power law of the form:

$$S_q(\ell) \propto \ell^{\zeta(q)} \tag{4}$$

where $\zeta(q)$ represents a *scaling function* depending on the order of the moment q. In Kolmogorov's theory, the energy spectrum of velocity signals for high-Reynolds flows follows a power law $\propto k^{-5/3}$ over a specific range of scales (inertial range) [7, 8]. Such a law can also be related by the structure function when $q = 2$ and the value of $\zeta(2)$ corresponds to 2/3, yielding a self-similarity scaling exponent equals to $H = \zeta(1) = 1 - \zeta(2) = 1/3$.

However, measurements and numerical experiments have determined that there exists a conditional self-similarity property in data since the structure function scaling exponents change for higher order of the moments, which could possibly be associated with the complex nature of intermittency in turbulent flows [3]. While a theory suggests that the energy dissipation in fully developed turbulent flows is concentrated in a *fractal set* whose dissipative structures have a *fractal dimension* that can only be determined in the sixth-order structure function and its value is near to the embedding dimension [3], another theory suggested that the explanation of such a behavior in the structure function is given by considering that there is a hierarchy of *singularities* of rates of dissipation ε of turbulent kinetic energy, which are distributed on interwoven sets of different fractal dimensions [5, 6].

The description of such a distribution can be obtained by thinking that the dissipation energy flux occurs in a domain of size λ_0 (or eddies of size λ_0), which is subdivided into two subdomains of size $\lambda_1 = \lambda_2 = \lambda_0/2$, but the dissipation energy flux is distributed in proportions p and $1 - p$ (with $p < 1$) for both resulting subdomains. The domain subdivision process is continued by applying the aforementioned distribution rule to each subdomain up to subdomains of size close to the Kolmogorov scale η, and the distribution of dissipation energy fluxes turns out by following the p-proportion rule to every sibling subdomains. At the end of the energy distribution process, there will be a structure statistically enriched by rates of dissipation ε of different sizes, which can be described by a hierarchy of singularities α with multiplicity of fractal dimensions $f(\alpha)$; thus, a multifractal description of the energy dissipation is suitable to explain the energy distribution and its statistical structure.

Turning to the study of rainfall, the use of the structure function exhibits some subtle differences. For rainfall time series, its structure function is defined in similar fashion as follows:

$$S_q(\tau) \equiv \langle |\mathrm{R}(t+\tau) - \mathrm{R}(t)|^q \rangle \equiv \langle |\Delta \mathrm{R}(\tau)|^q \rangle \tag{5}$$

where $\mathrm{R}(t)$ represents the rainfall field at time t, $\tau = nt/T$ (for $n = 1, 2, \dots$) is the aggregation timescale, and $\Delta\mathrm{R}(\tau)$ is the rainfall fluctuation for the aggregation timescale τ, and T is the observational period of time. Given the intrinsic relation between rainfall and turbulence, the expected rainfall's structure function follows a power law given by:

$$S_q(\tau) \propto \tau^{\xi(q)} \tag{6}$$

where $\xi(q)$ represents a scaling function depending on the order of the moment q. **Figure 1** shows an example of rainfall fluctuations time series for different

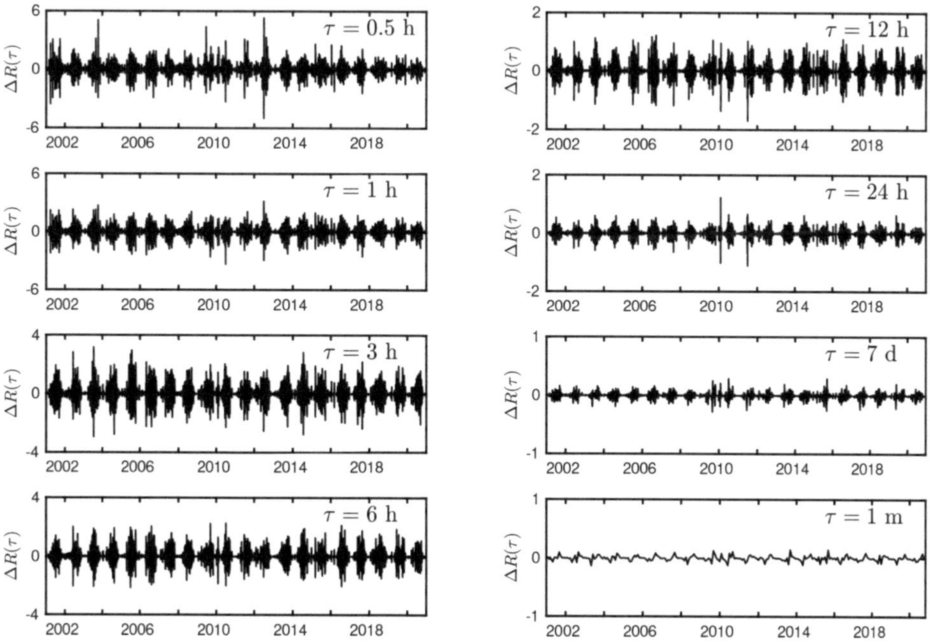

Figure 1.
Rainfall fluctuations time series. Each panel depicts rainfall fluctuations $\Delta R(\tau)$ for different timescales for the span period 2000 to 2020. The rainfall time series is derived from a spatial average of IMERG rainfall fields over the metropolitan area of Mexico valley on each time step. IMERG precipitation product has a time resolution of 30 minutes and spatial resolution of 0.1°. IMERG is the acronym of integrated multi-satellite retrievals for the global precipitation measurement mission.

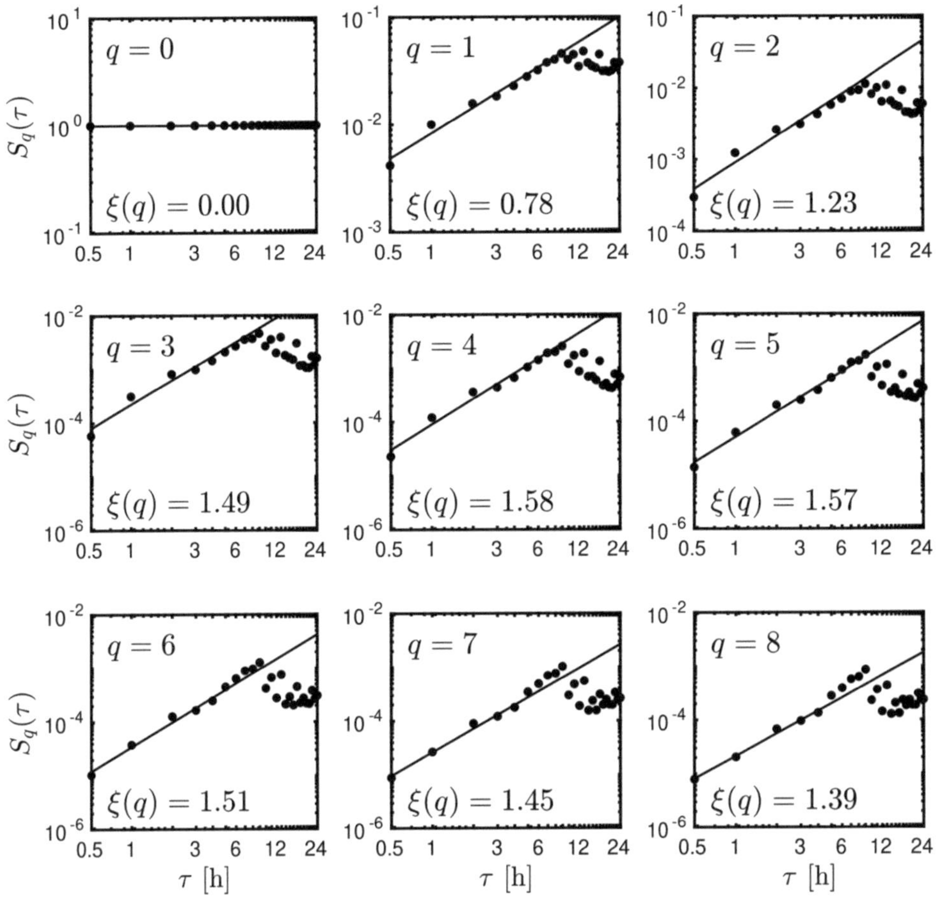

Figure 2.
Rainfall's structure function $S_q(\tau)$ estimations for several orders of the moments q. As seen, the data used for computing the structure functions corresponds to the rainfall fluctuations time series exhibited in **Figure 1**. *The rainfall's scale invariance mostly holds between the aggregation scales of $\tau = 30$ minutes to $\tau = 9$ hours. In this figure, the scaling exponent $\xi(q)$ of Eq. (6) is here obtained from the relation $S_q(\tau) \propto \tau^{\xi(q)}$ for the scale invariance range.*

aggregation timescales, whose data comes from IMERG satellite retrievals. One can easily appreciate in **Figure 1** rough changes in magnitudes and intermittency from scale to scale. **Figure 2** includes the corresponding structure function for the rainfall time series at aggregation timescales as shown in **Figure 1**. For $q > 0$, the scale invariance property is sustained for a noncompact set of aggregation timescales (i.e., from $\tau = 30$ min to $\tau = 9$ h); therefore, we could suggest that there is also a conditional self-similarity property in the statistical structure of rainfall due to its intrinsic intermittency. In the technical literature, it is suggested for the first-order structure function $S_1(\tau)$ that the scaling exponent $\xi(1)$ specifies how far the rainfall field is from a conserved field, and therefore, it indicates its order of fractional integration [9–12]. For the example shown in **Figure 2**, the scaling exponent values for the first four order structure functions are $\xi(1) \approx 0.8 \approx 4/5$, $\xi(2) \approx 1.2 \approx 6/5$, $\xi(3) \approx 1.4 \approx 7/5$, and $\xi(4) \approx 1.6 \approx 8/5$, whose scaling exponent values do not match the ones found in fully developed turbulence flows, that is, $\xi(1) = 1/3$, $\xi(2) = 2/3$, $\xi(3) = 3/3$, and $\xi(4) = 4/3$ [3, 4]. As seen in rainfall fluctuations, there seems to exist a conditional self-similarity property alike to the one identified in turbulence phenomena, which points out that the intermittency nature of rainfall patterns changes the symmetry property of the structure function.

Based on turbulence theory and its advances in the identification and characterization of multifractality, one can have some special considerations for rainfall analysis: (i) the physics of rainfall appears to be as complex as turbulence and hence physical models designed to describe rainfall physics and its scaling properties can either be implicitly implied in Navier-Stokes equations as a subprocess of the turbulence physics or be an independent physical process that interacts with the turbulence, and whose interactions have as consequence a tantamount scaling behavior; (ii) the observational evidence for rainfall shows that there exists a local scale-invariant property, which determines the existence of a noncompact set of scaling exponents [9–19], yielding the property of infinitely many scaling groups that characterizes strictly multifractal sets may not necessarily be satisfied in the same way for rainfall, therefore, suggesting the definition of its patterns as multiscaling sets.

2. Rainfall multiscaling properties

In a general form, a rainfall field $\mathrm{R}(t, \mathbf{x})$ at timescale λ_t and spatial scale λ_x will be formed by *measures* $\{\mu_i\}$ that possess multiscaling properties, that is, there exists a range of scaling (singularity) exponents $A_\alpha = (\alpha_{min}, \alpha_{max})$ defined on fractal sets with dimensions $f(\alpha)$. As the space or timescale is diminished ($\lambda \to 0$), the measures $\{\mu_i\}$ defined on $\mathrm{R}(t, \mathbf{x})$ should satisfy the following power-law relationships:

$$\mu_\alpha \propto \lambda^\alpha; \quad N(\alpha) \propto \lambda^{-f(\alpha)} \tag{7}$$

where the α-exponents are known as Hölder scaling singularity exponents of order α, $N(\alpha)$ is the number of elements with singularity exponents of order α, $f(\alpha)$ is the fractal dimension of the measure support with singularity exponents of order α, and λ represents the space (or time) scale of the rainfall field. The functional relationship between α and $f(\alpha)$ is known as the *multifractal spectrum* [1, 2, 20]. The former definition has a geometrical connotation as it implies the existence of a local singularity exponent α for every point of the measure function, which is not necessarily satisfied in observational data. Based on this restriction, the multiscaling properties of rainfall have also a *probabilistic* counterpart. If $g_X(X = \mu(\lambda)) \equiv g_X(\mu)$ is the probability density function (pdf) of the (random) measures $\{\mu_i(\lambda)\}$ defined on $\mathrm{R}(t, \mathbf{x})$ at the scale λ, the statistical moments of measures $\{\mu_i(\lambda)\}$ are given by:

$$M_\mu^q(\lambda) = \mathbb{E}[\mu^q(\lambda)] = \int_{\mathbb{R}} \mu^q(\lambda) g_X(\mu) d\mu \tag{8}$$

If Eq. 7 is satisfied for the rainfall measures set, the probability of having $\mu_\alpha \propto \lambda^\alpha$ should be proportional to $\lambda^{d-f(\alpha)}$. This is the probability of a ball at scale λ corresponding to a set C_α with dimension $f(\alpha)$ and Euclidean embedding dimension d. Therefore, the probability of having $\mu_\alpha \propto \lambda^\alpha$ can expressed as follows:

$$\lim_{\lambda \to 0} \frac{\log \mathbb{P}_r\{\mu > \lambda^\alpha\}}{\log \lambda} = d - f(\alpha) \tag{9}$$

This means that the integral in Eq. (8) can be redefined as:

$$M^q_{\mu_\alpha}(\lambda) = \int_{A_\alpha} \lambda^{q\,\alpha+d-f(\alpha)}\,d\mu_\alpha \equiv \int_{A_\alpha} \lambda^{q\,\alpha+c(\alpha)}\,d\mu_\alpha \tag{10}$$

where $c(\alpha) = d - f(\alpha)$ is the codimension function for the singularity exponents of order α. As suggested by the steepest-decent method, in the limit $\lambda \to 0$, the smallest exponents of the power law into the integral of Eq. (10) dominate; therefore, there exists a scaling exponent function $\theta_q = \inf_\alpha [q\,\alpha + d - f(\alpha)]$ that explains how the probability changes, this is:

$$M^q_\mu(\lambda) \propto \lambda^{\theta_q}; \quad \theta_q = \lim_{\lambda\to 0} \frac{\log M^q_\mu(\lambda)}{\log \lambda} \tag{11}$$

As such, the steepest-decent method points out that the integral of Eq. (10) is dominated by those terms where the integrand attains its maximum value, which also implies that:

$$\frac{d}{d\alpha}(f(\alpha) - q\alpha) = \frac{d\tau(q)}{d\alpha} = 0 \tag{12}$$

where $\tau(q) = f(\alpha(q)) - q\alpha(q)$ is defined as the *mass exponent* and represents a functional of three variables q, $\alpha(q)$, and $f(\alpha(q))$. Under this definition of the mass exponent $\tau(q)$, the singularity exponents of order α and the fractal dimension associated with those singularity exponents $f(\alpha)$ can be computed as follows:

$$\alpha = -\frac{d\tau(q)}{dq} \tag{13}$$

$$f(\alpha) = \tau(q) + q\alpha \tag{14}$$

Eqs. (13) and (14) represent what is called a *Legendre transformation*. In a similar manner, the scaling exponent function θ_q is also associated with a Legendre transformation of the form:

$$\theta_q = \inf_\alpha \{q\,\alpha + d - f(\alpha)\}; \quad \alpha = \frac{d\theta_q}{dq}; \quad f(\alpha) = \inf_\alpha \{q\,\alpha + d - \theta_q\}; \quad q = \frac{df(\alpha)}{d\alpha} \tag{15}$$

The function $f(\alpha)$ that describes the fractal dimension of singularities of order α turns out to be concave, and therefore, $d^2 f(\alpha)/d\alpha^2 \le 0$. Besides, for two singularity exponents $\{\alpha_i, \alpha_j\}$ with $\alpha_i < \alpha_j$, the concavity property suggests that for any value $\varepsilon \in (0, 1)$ the following inequality should be satisfied:

$$f((1-\varepsilon)\,\alpha_i + \varepsilon\,\alpha_j) \ge (1-\varepsilon) f(\alpha_i) + \varepsilon f(\alpha_j) \tag{16}$$

To summarize, $\mathbb{P}_r\{\mu > \lambda^\alpha\}$ as the probability distribution of the rainfall (random) measures $\{\mu_i(\lambda)\}$ at scale λ and also the geometric distribution of measures with singularities of order α and fractal dimension $f(\alpha)$. The previous definitions suggest that there is a power-law description of the rainfall measures statistics and a multiscaling process that shapes the rainfall geometric patterns. The application of these concepts have been extended to the space-time description of rainfall and the statistical characterization of rainfall events [14, 19, 21, 22]. So far, the technical literature reports a good fit of this multiscaling theory for understanding the statistical structure of rainfall and consequently the formely descriptive models of rainfall have been redefined to include multiscaling properties.

3. Multifractal approaches for modeling rainfall

In the last 30 years, several multifractal techniques and theories have been developed for describing and modeling rainfall [10, 17, 23–34]. The first approaches for such a task apply the phenomenology of multiplicative random cascades (e.g., [10, 14, 16, 26]). First, the so-called binomial random cascade will be studied next, as an application example and to introduce some properties of multiplicative random cascades.

Initially, one can suppose that there is a dyadic interval $I^k = I_{\beta_1\beta_2\ldots\beta_k}$ that is selected randomly and that such an interval represents the size of the simulated field and which helps to identify the location of simulated measures. Then a random sequence of digits $\beta_1\beta_2\ldots\beta_k$ taking values of 0 or 1 with a uniform probability distribution given by:

$$\Pr\{\beta_i\} = \begin{cases} 1/2, & \text{if} \quad \beta_i = 0 \\ 1/2, & \text{if} \quad \beta_i = 1 \end{cases} \qquad \text{for} \quad i = \{0,1,2,\ldots,k-1\} \quad \text{with} \quad k = 2^n \tag{17}$$

where $n = 0,1,2,\ldots$ may be used to determine the stages that define the random cascade with size determined by the dyadic interval I^k. The measures in such a dyadic interval are estimated following a multiplicative process, such that:

$$\mu\left(I^k\right) = \prod_{i=0}^{k-1} W_{\beta_i}; \qquad \text{with} \qquad W_{\beta_i} = \begin{pmatrix} W_0, & \text{if} \quad \beta_i = 0 \\ W_1, & \text{if} \quad \beta_i = 1 \end{pmatrix} \tag{18}$$

where W_{β_i} is a random variable whose possible values are W_0 and W_1 with probability 1/2, and the measure $\mu\left(I^k\right)$ is computed as products of k independent and identically distributed random variables W_{β_i}. The singularity exponents of order α associated with the measures and resulting from the multiplicative process are estimated as follows:

$$\alpha_k\left(I^k\right) = \frac{\log \prod_{i=0}^{k-1} W_{\beta_i}}{\log 2^{-k}} = -\frac{1}{k}\sum_{i=0}^{k-1} \log_2 W_{\beta_i} = \frac{1}{k}\sum_{j=1}^{k} V_j \tag{19}$$

where V_j is a random variable with probability distribution, which is given by:

$$\Pr\{V_j\} = \begin{cases} 1/2, & \text{for} \quad V_0 = -\log_2 W_0 \\ 1/2, & \text{for} \quad V_1 = -\log_2 W_1 \end{cases} \qquad \text{for} \quad j = \{1,2,\ldots,k\} \tag{20}$$

Hence, the singularity exponent α_k is also a random variable in which its expected value α_0 can be computed as:

$$\mathbb{E}[V_j] = \frac{1}{2}V_0 + \frac{1}{2}V_1 = -\frac{1}{2}[\log_2 W_0 + \log_2 W_1] = -\frac{1}{2}\log_2 W_0 W_1 \tag{21}$$

where $\mathbb{E}[V_j] \equiv \alpha_0$ represents the average singularity exponent to be found in a set of simulated measures. If one sets a restriction on the product of W_0 and W_1, such that its product is bounded in the range $0 \leq W_0 W_1 \leq 1$, the exponent α_0 will be singular (i.e., $0 \leq \alpha_0 \leq 1$) when the product is bounded in the range $1/4 \leq W_0 W_1 < 1$ and nonsingular (i.e., $\alpha_0 < 1$) when the product is bounded in the range $W_0 W_1 > 1/4$.

The singularity of a measures can also be understood through the cumulative measure function. At this stage, let us define the function $M(x) = \sum_{i=0}^{x} \mu(i)$ as a descriptor of the local variability of the field measures $\mu(i) = M(x(i) + \lambda) - M(x(i))$. With the help of Eq. (7), one can observe in the case when $\lambda = 2^{-k}$ and $k \to \infty$, if $\alpha > 1$ the measures $\mu \to 0$ and the function $M(x)$ becomes constant for all values of x (nonsingular measures), but if $0 \leq \alpha \leq 1$ the measures $\mu_i \to 1$ and the function $M(x)$ characterizes the measures as *singular* for all values of x. Therefore, the values to be selected for W_0 and W_1 will determine the singularity of the binomial random cascade measures.

Now, the variability of binomial random cascade measures can be approximated by the variance of α_k, thus:

$$\mathrm{Var}[V_j] = \mathbb{E}\left[V_j^2\right] - (\mathbb{E}[V_j])^2 \tag{22}$$

$$= \left(\frac{1}{2}V_0^2 + \frac{1}{2}V_1^2\right) - \left(\frac{1}{2}V_0 + \frac{1}{2}V_1\right)^2 = \frac{1}{4}(V_0 - V_1)^2 \tag{23}$$

$$= \frac{1}{4}\left(\log_2\left(\frac{W_0}{W_1}\right)\right)^2 \tag{24}$$

and the expected limits of the singularity exponent α are defined by:

$$\alpha_1 = -\frac{1}{2}\left(\log_2 W_0 W_1 + \log_2\left(\frac{W_0}{W_1}\right)\right) = -\frac{1}{2}\log_2 W_0 \tag{25}$$

$$\alpha_2 = -\frac{1}{2}\left(\log_2 W_0 W_1 - \log_2\left(\frac{W_0}{W_1}\right)\right) = -\frac{1}{2}\log_2 W_1 \tag{26}$$

where $\alpha_1 = \alpha_{min}$ and $\alpha_2 = \alpha_{max}$ if $W_0 > W_1$. On the other hand, if the *strong law of large numbers* (SLLN) is applied to Eq. (19), the expected value of the singularity exponent α_k will converge, almost surely, to the expected value α_0 for $k \to \infty$, that is,

$$\mathbb{P}_r\left\{\lim_{k\to\infty}\frac{1}{k}\sum_{j=1}^{k} V_j = \alpha_0\right\} = 1 \tag{27}$$

The previous result means that the binary expansion $\beta_1\beta_2 \ldots \beta_k$ will have the same frequency of zeros and ones with probability 1. Moreover, for a large number of finite size 2^k, there will be a range of singularity exponents spanning from α_{min} to α_{max}; thus, deviations from the expected value α_0 become significant depending on the value of k. As one can observe, the binomial random cascade is a model for describing multiscaling processes where its patterns are enriched with singularity exponents of order α but defined as sums of independent and identically distributed random variables W_{β_i}.

As mentioned above, the binomial random cascade approach has been used for describing the distribution of the energy dissipation fluxes in turbulent flows [5, 6]; however, its application for the description of rainfall field measures has not been completely successful. In observational rainfall data, there is not a well-defined compact and dense set of singularity exponents which possibly is due to measurements problems, insufficient length of rainfall records, or the intrinsic aperiodic structure of their patterns.

The next models to be presented exhibit considerations for modeling rainfall under a multiscaling approach. There are always limits for a complete description, and caution must be taken at the moment of selecting the model for particular applications. Let us begin with one pioneering multiscaling (or multifractal) rainfall model, which was introduced by Over and Gupta [16], and is set up on the theory of random cascades. In this model, rainfall fields are based on the subdivisions of a d-dimensional domain $[0, L_0]^d$ into b^n subcubes. For instance, if the embedding dimension equals $d = 2$ and the branching number equals $b = 4$, there are 4^n subdivisions of the two-dimensional rainfall field at the n-th level of the cascade. As such, the i-th subdomain after n partitions is denoted by Δ_n^i and the length $L_n = L_0 b^{-n/2}$ of each side of the subdomain Δ_n^i at the n-th level represents the spatial scale.

The construction of the rainfall field begins when a domain $[0, L_0]^d$ adopts an initial rainfall mass $R_0 L_0^d$, and the i-th subdomain of the first subdivision Δ_1^i consequently takes a rainfall mass equals to $R_0 L_0^d W_1^i b^{-1}$ for $i = \{1, \dots, b\}$. Here, W_1^i is a mutually independent random variable and the distribution of W is called the cascade generator. Thus, for every subdivision Δ_n^i of the cascade, the mass is estimated as follows:

$$\mu\left(\Delta_n^i\right) = \frac{R_0 L_0^d}{b^n} \prod_{j=1}^{n} W_j^i \tag{28}$$

Following Eq. (28), the cascade limit mass μ_∞ is obtained by letting $n \to \infty$ (small-scale limit), so that the rainfall field is computed as:

$$\mu_\infty = \mu(\Delta_n) Z_\infty \tag{29}$$

where,

$$Z_\infty = \frac{\mu_\infty\left([0, L_0]^d\right)}{R_0 L_0^d} \tag{30}$$

is an independent and identically distributed set of random variables which is employed to designate the high-frequency (or small-scale) component of the rainfall field. Z_∞ can also be understood as a normalized mass of the rainfall field when the multiplicative process is developed for the subdivision $n \to \infty$. The low-frequency (or large scale) component of the rainfall field is designated by $\mu(\Delta_n)$, and thus, the final construction of the random cascade is defined by the product of both high- and low-frequency components. Another important aspect in the construction of random cascade fields is their ability to maintain the ensemble mean of the random variable W

which is defined as $\mathbb{E}[W] = 1$, thus preserving mass on the average at all scales for each partition of the random cascade. It ought be highlighted that this approach may yield degenerate cascades, that is, those with total mass μ_∞ of zero with probability of 1. For a nondegenerate cascade, a suitable cascade generator should be carefully selected and for all cases, and the canonical restriction $\mathbb{E}[W] = 1$ and $\mathbb{E}[Z_\infty] = 1$ should be satisfied.

An example of a possible outcome from this multifractal model is illustrated in **Figure 3**. An advantage of this random cascade model is that it is parsimonious, which means that it requires few parameters to describe the spatial pattern. However, the model foundations are based on the theory of random functions and, for this reason, there are some difficulties in validating the choice of a universal class of random generators for a particular case of study. On the other hand, it ought to be highlighted that a common drawback of random cascade models is that their simulations often exhibit what is called blockiness, that is, a concentration of the mass in blocks into the spatial patterns [26, 35]. Although there exist several procedures for vanishing the anomalous effect of blockiness in random cascade outputs, further research is required for improving the reduction of blockiness from the model itself.

Having described a probabilistic approach to represent rainfall fields, here we devote our attention to another model, which defines multifractal measures but via a deterministic approach. This is the so-called *Fractal-Multifractal* (FM) *approach* that was introduced by Puente [32] and has been applied to rainfall modeling by several researchers in the last 30 years, for example, [23, 25, 29, 31, 33, 34, 36]. In this model, the observed patterns are built as illuminated canonical deterministic multifractal measures that are supported over the graphs of *fractal interpolation function* [37].

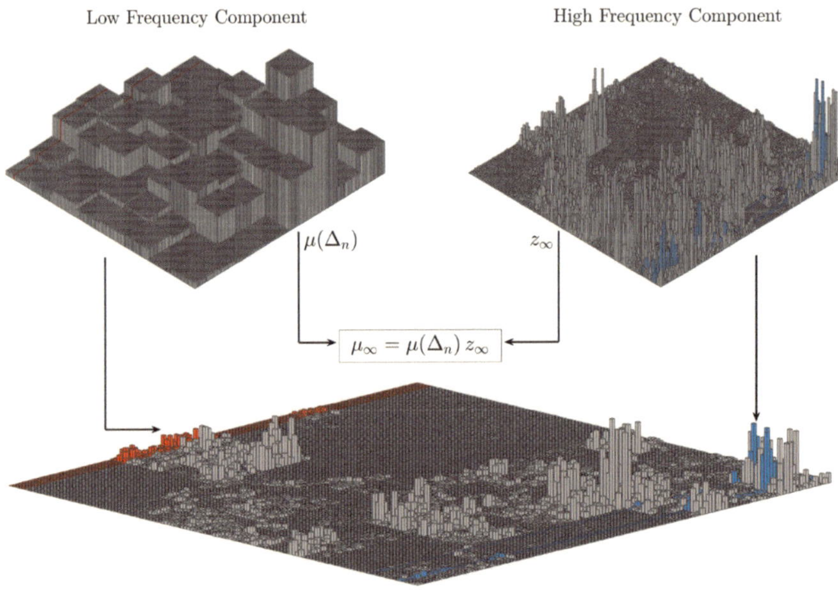

Figure 3.
Random cascade field simulated with the algorithm suggested by reference [16]. The field was simulated for an unitary precipitation rate and a spatial domain of 128 × 128 spatial units. The model used a branching number of 4, the subdivisions for the high-frequency component (z_∞) were seven stages and the subdivisions for the low-frequency component ($\mu(\Delta_n)$) were three stages.

In the original FM approach, a combination of linear functions of the form $W_n(\mathbf{x}) = A_n \mathbf{x} + B_n$ (*affine transformations*) are employed to interpolate (in a fractal way) a set of points in $\mathbb{R}^N$ and then, by using projections (derived distributions) of canonical multifractal measures over the graphs of such functions, other, noncanonical, multifractal measures are constructed. For instance, at $\mathbb{R}^2$, the FM approach requires a set of at least two contractile affine maps for modeling patterns along a line, that is,

$$W_n\begin{pmatrix} x \\ y \end{pmatrix} = \begin{pmatrix} a_n & 0 \\ c_n & d_n \end{pmatrix}\begin{pmatrix} x \\ y \end{pmatrix} + \begin{pmatrix} e_n \\ f_n \end{pmatrix} = Ax + t, \quad n = 1, \dots, N \tag{31}$$

subject to these restrains:

$$W_n\begin{pmatrix} x_0 \\ y_0 \end{pmatrix} = \begin{pmatrix} x_{n-1} \\ y_{n-1} \end{pmatrix}, \quad W_n\begin{pmatrix} x_N \\ y_N \end{pmatrix} = \begin{pmatrix} x_n \\ y_n \end{pmatrix}, \quad 0 \leq d_n < 1, \quad n = 1, \dots, N \tag{32}$$

where a_n, c_n, e_n, and f_n are parameters of the FM approach in $\mathbb{R}^2$, which are estimated directly from the restrains indicated above. d_n is also a parameter of the FM approach, but this one defines the vertical scaling factor in the transformation W_n. Consequently, all affine transformations $W_n(\cdot)$ become contractile if for all n is satisfied that $0 \leq d_n < 1$. When all these conditions are satisfied, a unique fixed point exists, that is, a fractal interpolation function $f : x \rightarrow y$, such that $G = \{(x, f(x)) | x \in [0, 1]\}$, that satisfies $G = \cup_{n=1}^{N} W_n(G)$ [37]. During the construction of the fractal interpolation function in $\mathbb{R}^2$, two probability measures dy and dx are generated by counting the frequency of points within coordinates x and y. Under a suitable parameter configuration, both generated measures are multifractal objects: the one over x is a canonical multinomial multifractal and the one over y may resemble geometrically geophysical patterns [33]. In addition, if for all points at coordinates x and y a rotation θ is applied, the original FM approach increases its complexity, that is, enhances its space of possibilities for generating yet more interesting fractal patterns. The concept of projections of multifractal measures that are supported over a fractal interpolation function are indeed transformations $W_n(x) : \mathbb{R}^2 \rightarrow \mathbb{R}^2$, where the matrix A_n in $W_n(x) = A_n x + t_n$ is rewritten as follows:

$$A_n = \begin{pmatrix} a_n & 0 \\ c_n & d_n \end{pmatrix} \rightarrow A_n = \begin{pmatrix} \cos\theta & -\sin\theta \\ \sin\theta & \cos\theta \end{pmatrix}\begin{pmatrix} a_n & 0 \\ c_n & d_n \end{pmatrix}^T \tag{33}$$

where the matrix A_n transforms the relative space with rotations or contractions, and the vector t_n specify a linear translational displacement of the map. The new measures under that comes out from the rotational transformation are now denotated as $dy|_\theta$ and $dx|_\theta$. **Figure 4** illustrates some simulated rainfall fields via FM approach in $\mathbb{R}^2$. While the upper left graph shows a fractal interpolation function and over it the corresponding measures obtained from a binomial cascade, the one on the upper right shows the projection (or illumination) of those multifractal measures for a specific rotation angle θ. The lower left graph of **Figure 4** shows the comparison between a simulated rainfall field and the high-resolution rainfall pattern registered at Boston in Octuber 25, 1980 (the right plot) which was previously studied by Puente and Obregón [33]. Some applications of the FM approach for modeling spatial patterns can be found in Refs. [23, 33].

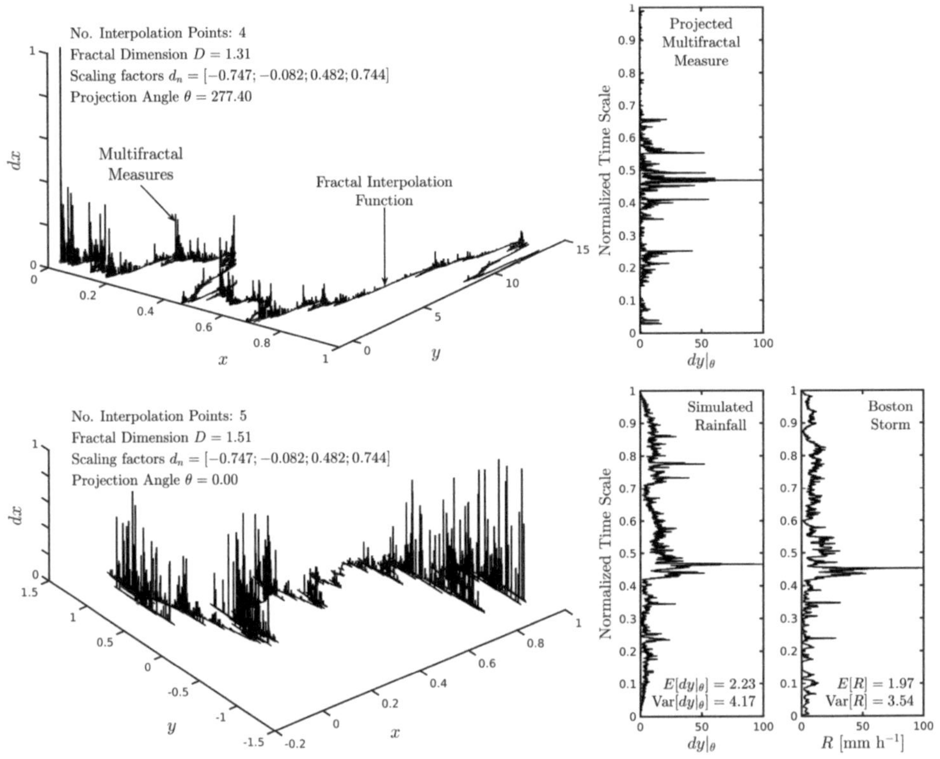

Figure 4.
Example of two simulated precipitation fields via the FM approach. In each subfigure, the left panel shows the construction of a fractal interpolation function in the plane x–y and over such a function, multifractal measures dx are exhibited. The projection concept of multifractal measures are represented by the model as a new derived measure $dy|_\theta$ (second panel), which resemble complex geophysical patterns. In the lower subfigure is depicted an output of the FM approach that was originally reported in ref. [33], but adapted in this manuscript by the authors who used an own MATLAB algorithm. The simulated rainfall resembles graphically and statistically the rainfall pattern observed in Boston on October 25, 1980 (lower left panel).

4. Space: time multifractal modeling

There are many applications on meteorology and hydrology in which a good knowledge about the dynamics of the space-time precipitation field would be very useful to know in advance (e.g., early warning systems for flood control). Although there exists many climatic and weather models, trying to obtain prediction to be used as inputs of rainfall-runoff models, there are difficulties to be compatible with the scales among these models. If the output scale from precipitation forecasting models is higher than the necessary scale for running a hydrological model, then it is required to employ a downscaling procedure for parity among data scales, but such a procedure entails to information losses and a reduction of the forecasting likelihood. Clearly, the relationship between scales that are used for running weather models and those ones used for running hydrological models is still a problem to be solved.

In the multiscaling theory of the rainfall field, a self-similar process is assumed to explain the space-time dynamics of rainfall, that is, there are sequences of precipitation data that converge in law after a change of location and timescale. Formally, a process $\mathrm{R}(t, \mathbf{x})$ on the interval $\mathbb{R}^+ = [0, \infty)$ is self-similar with exponent H if for any $c > 0$ and $d > 0$, the following property is satisfied:

DOI: http://dx.doi.org/10.5772/intechopen.1001894

$$\{\mathrm{R}(ct, d\mathbf{x})\} \stackrel{\mathrm{d}}{=} \left\{(cd)^H \mathrm{R}(t, \mathbf{x})\right\} \tag{34}$$

In addition, the self-similarity property represents also the validity of the Taylor's hypothesis of the "frozen turbulence" which says that the autocovariance at a time lag $\tau = t_2 - t_1$ (with $t_2 > t_1$), and at a fixed but arbitrary location x_0, is the same to the spatial covariance between two spatial points separated by the translation distance $u\tau$ for at a fixed but arbitrary time t and a storm moving at a constant velocity u, this is:

$$\mathrm{Cov}[\mathrm{R}(t_1, x_0), \mathrm{R}(t_2, x_0)] = \mathrm{Cov}[\mathrm{R}(t, x_0), \mathrm{R}(t, u\tau)] \tag{35}$$

In general form, Taylor's hypothesis reinterprets time variation at a fixed point as a spatial variation at some fixed time; besides, Taylor's hypothesis represents a space-time process which is forced by a constant large-scale advection velocity $u = \lambda/\tau$ at the spatial scale λ. Such advection velocity u satisfies the scaling transformation property $u \to \lambda^H u$, which was indicated previously in Eq. (2) and H represents a self-similarity scaling exponent.

It is highlighted that there are some physical considerations in the application of Taylor's hypothesis. Some empirical studies have reported that the large-scale advection velocity induces a space-time anisotropy (e.g., [29, 38]). As the advection velocity changes with time, the anisotropy also changes; therefore, there are limit values for the time lag τ (measured from the origin of the storm) for the Taylor's hypothesis to be satisfied. Furthermore, there could be similar near-zero velocities in the rain cells or in the storm region that also break the hypothesis.

Several studies have proved that time and space scaling properties can be represented by power laws in their structure functions such that:

$$S_q(\tau) \propto \tau^{H_\lambda(q)} \qquad (\text{for a local area of size} \lambda) \tag{36}$$

$$S_q(\lambda) \propto \lambda^{H_\tau(q)} \qquad (\text{for a duration } \tau) \tag{37}$$

In practice, both space and time structure functions are restricted to some temporal and spatial scales; therefore, the scaling invariance property is bounded (i.e., there is a range in which the multifractal scaling properties are broken or change to represent a monofractal behavior) and such a consideration sets limits in the application of rainfall models. Another problem is the identification of scaling relationships between the exponents $H_\lambda(q)$ and $H_\tau(q)$ or to identify a new one that does not depend directly on either spatial or timescales. Deidda [38] suggests that rainfall can be modeled as a self-similar multifractal process (for a bounded spatial range $\lambda_0 \leq \lambda \leq L$) with a constant advection velocity that satisfies the Taylor's hypothesis. Moreover, the space-time structure function:

$$S_q(\tau, \lambda) = \langle |\mathrm{R}(t + \tau, \mathbf{x} + \lambda) - \mathrm{R}(t, \mathbf{x})|^q \rangle \tag{38}$$

is given by the following space-time scaling law:

$$S_q(\tau, \lambda) \propto \lambda^{K(q)} \tag{39}$$

where the space-time scaling exponent function $K(q)$ shown here does not dependent on either time or space scales in a direct form. If one considers only the space-time anisotropy due to the large-scale advection velocity, the time domain can be rescaled with the advection velocity in order to maintain the space-time self-similarity property.

In Deidda's approach [38], there exists a scale-dependent velocity parameter u_λ which is approximated as λ^H, where H is the anisotropic scaling exponent. If $H = 0$, the velocity parameter is constant and independent of the scale λ, if $H > 0$ the velocity parameter increases at larger scales and conversely, if $H < 0$ the velocity parameter increases at smaller scales. For a multiplicative random cascade process with branching number b, the spatial scale would be given by $\lambda_n = Lb^{-n}$ (for $n = [0, \infty)$), the scale-dependent velocity parameter would be given by $u_\lambda = U_0 b^{-nH}$, and the timescale would be $\tau_n = \lambda_n u_\lambda^{-1} = LU_0^{-1} b^{-n(1-H)}$.

Deidda [38] also suggest that to estimate the anisotropic scaling, exponent is necessary to satisfy the self-affinity property which is given by:

$$\frac{P(b_n\omega)}{P(\omega)} = \frac{P(b_n k_x)}{P(k_x)} = \frac{P(b_n k_y)}{P(k_y)} \tag{40}$$

defined by the following power laws:

$$P(\omega) \propto \omega^{-s_t}; \quad P(k_x) \propto k_x^{-s_x}; \quad P(k_y) \propto k_y^{-s_y} \tag{41}$$

where $P(\cdot)$ represents the power spectrum, ω is the frequency, k_x and k_y are the wave numbers in the zonal and meridional directions, respectively, and s_x, s_y, and s_t represent the slopes of the space-time power spectrum. The zonal anisotropic scaling exponent is then computed as $H_x = 1 - s_x/s_t$ and the meridional anisotropic scaling exponent as $H_y = 1 - s_y/s_t$.

On another point of view, the aforementioned self-affinity property can be understood as an space-time transformation such that:

$$\mathbf{x} \to \frac{\mathbf{x}}{b_s}; \qquad t \to \frac{t}{b_t} \tag{42}$$

where b_s identifies the branching number for space and b_t is the branching number for time. Since the space and time branching numbers are related through the anisotropic scaling exponent H, the time branching number can be estimated as $b_t = b_s^{1-H}$.

In similar fashion to Deidda [38], Marzan et al. [28] have proposed a space-time multifractal model based on the concept of anisotropic scaling and the existence of a causal process in the space-time dynamics of rainfall (i.e., the past influences the future). In this model, rainfall intensity values $\mathrm{R}(t, \mathbf{x})$ represent random spatial structures of scale λ and density $\varepsilon(\mathbf{x})$. For a discrete random cascade model, densities evolve by computing products between random variables as follows:

$$\varepsilon_{n+1}(\mathbf{x}) = \varepsilon_n(\mathbf{x}) W_{n+1}(\mathbf{x}) \tag{43}$$

where W is a random random variable that satisfies the scaling law $\langle W^q \rangle \approx \lambda^{K(q)}$ and the positivity property $\mathbb{P}_r\{W < 0\} = 0$. At each step of the cascade, the structures have a scale given by $\lambda_n = Lb^{-n}$, which become structures of scale $\lambda_{n+1} = Lb^{-(n+1)}$ and so on. In the space-time domain, the evolution of scales is given by $\lambda_{n+1} = Lb^{-(n+1)}$ for the spatial domain and $\tau_{n+1} = Tb^{-(n+1)(1-H)}$ for the time domain. Under this conceptual model, if the spatial branching number equals to $b_s = 8$ and the time branching number equals to $b_t = 4$, the anisotropic scaling exponent should be equal to $H = 1/3$, which is the value expected for passive scalar fields in turbulent environments. Clearly, one can observe that the lifetime of structures depend on the spatial scale and the degree of space-time anisotropy. For $H \to 0$, structures change at the same velocity either time or space, but for $H \to 1$, structures evolve faster in time than in the space. **Figure 5** shows a discrete space-time simulation of the rainfall field by following the conceptual model suggested by Marzan et al. [28]. In this figure, the space is represented in a one-dimensional domain and so is the time; the rainfall intensity measures turn out from a random cascade simulation run, which evolves in time according to Eq. 43 and accounting an anisotropy scaling exponent $H \approx 1/3$.

For generating continuous cascades, the multiplicative process can be replaced by an additive process of the logarithms of field values. Initially, the cascade generator is given by Γ_λ at the scale λ, then the densities of the field are computed as:

$$\varepsilon_\lambda(t, \mathbf{x}) = e^{\Gamma_\lambda(t,\mathbf{x})} \tag{44}$$

Once again, the field will be multifractal if the scaling law $\langle \varepsilon_\lambda^q \rangle \approx \lambda^{K(q)}$ or equivalently, $\langle e^{q\Gamma_\lambda} \rangle \approx e^{K(q)\log \lambda}$. Besides, the multifractal field generator can be expressed as an impulse-response function between a scale function $g_\lambda(t, \mathbf{x}) \approx \|(t, \mathbf{x})\|^h$ and a random function $\gamma_\lambda(t, \mathbf{x})$:

$$\Gamma_\lambda(t, \mathbf{x}) = g_\lambda(t, \mathbf{x}) * \gamma_\lambda(t, \mathbf{x}) = \int_0^t g_\lambda(\tau, \mathbf{x})\gamma_\lambda(t + \tau, \mathbf{x})d\tau \tag{45}$$

where this integration represents a sum of random variables at different scales. The random function $\gamma_\lambda(\mathbf{x}, t)$ in Eq. (45) represents a causal filter, which can be expressed as follows:

$$g_\lambda^{-1}(\mathbf{x}, t) * \Gamma_\lambda(t, \mathbf{x}) = \gamma_\lambda(\mathbf{x}, t) \tag{46}$$

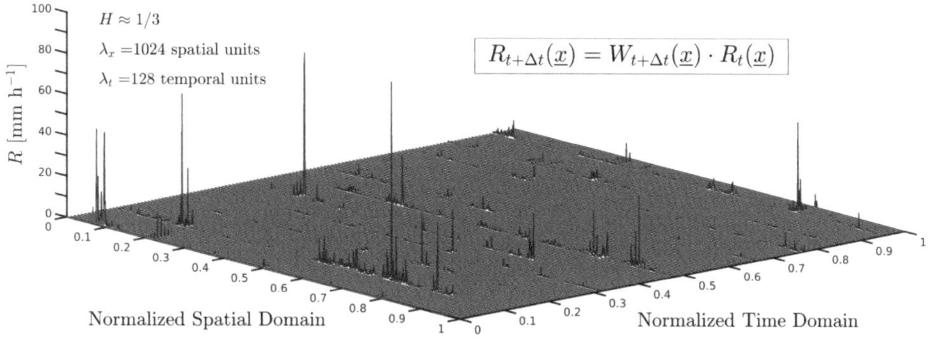

Figure 5.
Example of a discrete space-time random cascade. This construction uses an anisotropic scaling exponent $H \approx 1/3$, a branching number $b = 2$ and nine stages of the cascade development. The space and time domains were normalized between 0 and 1; however, the spatial domain is discretized by 1024 bins and the time domain by 128. The generator of the random variable W used in this realization a standard normal distribution. The time changes of the 1-D spatial rainfall field evolve according to Eq. (43).

where $g_\lambda(\mathbf{x}, t)$ represents a Green function and that corresponds to a fractional diffusion equation of the form:

$$\left(\partial_t^{\xi_1} - \partial_{\mathbf{x}}^{\xi_2}\right)\Gamma_\lambda(\mathbf{x}, t) = \gamma_\lambda(\mathbf{x}, t) \quad (47)$$

with,

$$\xi_1 = \frac{h(1+1/d)}{(1-H)}; \qquad \xi_2 = \frac{h(1+1/d)}{2}; \qquad \hat{g}_\lambda(\omega, |k|) = \frac{1}{|\omega|^{2\xi_2} - (i\omega)^{\xi_1}} \quad (48)$$

where d is the Euclidean spatial dimension and H is the anisotropic scaling exponent. Eq. (47) suggests the necessity of of employing an anisotropic scaling to represent the space-time asymmetry observed in rainfall patterns. In order to describe mathematically the anisotropic scaling, a fractional differential equation is suggested for the multifractal field generator; furthermore, the well-known multiplicative process used for describing spatial fields is changed by an additive process which constitutes a mechanism of modeling random cascades as a fractional diffusive process.

Another approach for the space-time modeling, besides the ones mentioned above, and one which considers explicitly an anisotropy between time and space, was introduced by Over and Gupta [30]. In this approach, the spatial precipitation fields are modeled as random multiplicative cascades that change in time by considering that their cascade generators can be represented as a time-indexed stochastic process. Under such a consideration, spatial precipitation fields evolve by following a evolutionary causal stochastic process. Moreover, the approach is settled in a discrete theory instead that a continuous one since it eases the estimation of nonrainy areas and its application for solving some hydrologic problems.

In the previous section, the Over and Gupta's approach [16] was presented for describing the construction of random cascade fields without a time indexation. The extension of this approach for describing space-time precipitation fields implies as mentioned above that the cascade generator of W be replaced by an independent and identically distributed stochastic process $\{W_t\}$ indexed to time, which should satisfy for each time step:

$$\mathbb{P}_r\{W_t \geq 0\} = 1; \qquad \mathbb{E}[W_t] = 1 \quad (49)$$

The cascade limit mass at time t is defined as $\mu_\infty\left(\Delta_n^i, t\right)$ with a cascade generator W_t, which possesses a temporal cross moment defined by:

$$M_n^q(t_1, t_2) = \sum_{i=1}^{b^n} \mu_\infty^q\left(\Delta_n^i, t_1\right)\mu_\infty^q\left(\Delta_n^i, t_2\right) \quad (50)$$

Since the field masses are identically distributed and they obeyed the law of large numbers, the ensemble moments are computed as the expectation of the temporal cross moment, thus:

$$\mathbb{E}\left[M_n^q(t_1, t_2)\right] = \left(\frac{R_0(t_1)R_0(t_2)L_0^{2d}}{b^{2n}}\right)^q \left(b^2\mathbb{E}\left[W_{t_1}^q W_{t_2}^q\right]\right)^n \mathbb{E}\left[Z_\infty^q(t_1)Z_\infty^q(t_2)\right] \quad (51)$$

Taking b-base logarithms to this equation gives:

$$\begin{aligned}\log_b \mathbb{E}\left[M_n^q(t_1,t_2)\right] = q \log_b\left(R_0(t_1)R_0(t_2)L_0^{2d}\right) + \log_b\left(\mathbb{E}\left[Z_\infty^q(t_1)Z_\infty^q(t_2)\right]\right) \\ +n\left(\log_b\left(\mathbb{E}\left[W_{t_1}^q W_{t_2}^q\right]\right) + 2(1-q)\right)\end{aligned} \tag{52}$$

or equivalently:

$$\begin{aligned}\log_b \mathbb{E}\left[M_n^q(t_1,t_2)\right] = q \log_b\left(R_0(t_1)R_0(t_2)L_0^{2d}\right) + \log_b\left(\mathbb{E}\left[Z_\infty^q(t_1)Z_\infty^q(t_2)\right]\right) \\ -2\log_b\left(\frac{L_n}{L_0}\right)\left(\log_b\left(\mathbb{E}\left[W_{t_1}^q W_{t_2}^q\right]\right) + 2(1-q)\right)\end{aligned} \tag{53}$$

From which the temporal cross moment can rewritten as follows:

$$\log_b M_n^q(t_1,t_2) = \frac{\log_b\left(M_n^q(t_1,t_2)\right)}{\log_b\left(\mathbb{E}\left[M_n^q(t_1,t_2)\right]\right)} + \log_b\left(\mathbb{E}\left[M_n^q(t_1,t_2)\right]\right) \tag{54}$$

$$= \log_b\left(Y_n^q(t_1,t_2)\right) + \log_b\left(\mathbb{E}\left[M_n^q(t_1,t_2)\right]\right) \tag{55}$$

where $Y_n^q(t_1,t_2)$ is a positive martingale sequence allowing the temporal cross moment to be approximated as follows:

$$\begin{aligned}\log_b M_n^q(t_1,t_2) \approx \log_b\left(Y_n^q(t_1,t_2)\right) + \log_b\left(\mathbb{E}\left[Z_\infty^q(t_1)Z_\infty^q(t_2)\right]\right) - \\ 2\log_b\left(\frac{L_n}{L_0}\right)\log_b\left(\mathbb{E}\left[W_{t_1}^q W_{t_2}^q\right]\right)\end{aligned} \tag{56}$$

This equation shows that for $L_n/L_0 \to 0$, the temporal cross moments of a single space-time realization are scale-invariant. Moreover, the stochastic definition of the random variable W_t determines the structure and variability of the temporal cross moments of the space-time rainfall field. For instance, if W_t is a time stationary stochastic process, then:

$$\begin{aligned}\log_b M_n^q(t_1,t_2) \approx \log_b\left(Y_n^q(t_1,t_2)\right) + \log_b\left(\mathbb{E}\left[Z_\infty^q(t_1)Z_\infty^q(t_2)\right]\right) - \\ 2\log_b\left(\frac{L_n}{L_0}\right)\log_b\left(\mathbb{E}\left[W_{t_1}^q\right]\right)\log_b\left(\mathbb{E}\left[W_{t_2}^q\right]\right)\end{aligned} \tag{57}$$

Now, if $\mathbb{E}\left[W_t^q\right]$ possess a power law distribution given by $\mathbb{E}\left[W_t^q\right] \approx (L_n/L_0)^{\theta_q}$, then:

$$\log_b M_n^q(t_1,t_2) \approx \log_b\left(Y_n^q(t_1,t_2)\right) + \log_b\left(\mathbb{E}\left[Z_\infty^q(t_1)Z_\infty^q(t_2)\right]\right) - 2\theta_q^2\left(\log_b\left(\frac{L_n}{L_0}\right)\right)^3 \tag{58}$$

where θ_q is a space-time moment scaling exponent. If in addition the time is related to the spatial scale via an anisotropic scaling exponent H, as suggested by [28]:

$$\frac{L_n}{L_0} = (t_2 - t_1)^{1/(1-H)} \tag{59}$$

The temporal cross moment would be given by the following expression:

$$\log_b M_n^q(t_1, t_2) \approx \log_b\left(Y_n^q(t_1, t_2)\right) + \log_b\left(\mathbb{E}\left[Z_\infty^q(t_1) Z_\infty^q(t_2)\right]\right) - 2\theta_q^2 \left(\frac{\log_b(t_2 - t_1)}{1 - H}\right)^3 \tag{60}$$

The last equation suggests that the divergence of temporal cross moments is set for an anisotropic scaling exponent close to unity (i.e., $H \rightarrow 1$); hence, a well-defined algebraic structure of the space-time scaling moments requires that $H > 1$.

Clearly, Over and Gupta's approach [30] and the others mentioned above have some limitations to be used as a general theory of the space-time structure of rainfall fields. Certainly, a key aspect in the definition of a suitable rainfall model for a specific region, besides of having an anisotropic scaling exponent, is the elucidation of a Markovian process for the random cascade generator in order to arrive at a causal process. Further works should be addressed to structure an overall Markovian process to represents the space-time nature of rainfall dynamics, hopefully, in connection with a physical atmospheric theory.

5. Summary and conclusions

This chapter highlighted that the multiscaling approach is a landmark theory for the description of the space-time structure of rainfall, crucial for improving the current research and applications in meteorology and hydrological sciences. Clearly, the usage of such schemes in the world depends on the advances in the interpretation and prediction of upcoming climatic events, even more in those regions with scarcity of meteorological observations or where the climatic data are necessarily reconstructions of models outcomes.

One of the ideas highlighted in this chapter is that some complex natural phenomena including turbulence can be described by multiscaling properties. In a similar fashion, the space-time structure of rainfall can be described by multiscaling properties, which enlighten the complex composition of the rainfall field in terms of singular measures, self-similar groups, and power laws. As explained in the chapter, the identification of multiscaling properties in the rainfall fields relies on adopting and adapting the mathematical techniques that were originally applied in turbulence theory. Although rainfall and turbulence are interwoven by similar physical processes, the statistical and geometrical properties vary between them; moreover, the concept of universality in turbulence does not seem to hold on rainfall, because many studies have reported changes in rainfall scaling properties depending on regional or local conditions. The aforementioned observation suggests that rainfall is an inherently complex natural process that requires being understood locally; hence, rainfall models should be designed to specific regional conditions and attached to the available measurement techniques.

As shown here, fractal theories represent the cornerstone in the formulation of a theory for the description of space-time rainfall processes. The predominant characteristics of rainfall statistics are indeed given by power laws, singular scaling exponents, and fractal dimensions for the set of rainfall measures. Our understanding of such characteristics have allowed the birth of several models that are settled in stochastic and geometrical theories. The multiscaling models for rainfall that were mentioned herein can describe space or time patterns with a good resemblance of those ones found in observational fields; however, these models exhibit limitations that prevent their usage under general conditions, specially in their adaptation to the space-time analysis.

In rainfall, time and space models are supposed to be linked via an isotropic scaling exponent, which in turn represents a symmetry breaking of the time-space domain. On the other hand, the most acknowledged space-time multifractal (or multiscaling) conceptualization recognizes the necessity of including a causal process for the description of rainfall field temporal changes. Under the basis of the linear theory, Marzan et al. [28] proposed a fractional diffusive approach for getting a causal process, but the Over and Gupta's approach [30] suggests the usage of a time-indexed stochastic process. Although these approaches exhibit some conceptual parallelism on their theoretical formulation, there still exits differences that preclude defining a unified theory. Moreover, their applications in general regions require an appropriate knowledge of the space-time scaling structure of the rainfall field in order to correctly define their parameters.

Acknowledgements

The authors thank all the technical support provided by the Institute of Atmospheric Science and Climate Change (ICAyCC)—Universidad Nacional Autonóma de México (UNAM). The first author thanks the General Direction of Academic Staff Affairs (DGAPA—UNAM) for its postdoctoral scholarship.

Conflict of interest

The authors declare no conflict of interest.

Nomenclature

α	singularity scaling exponent
α_0	average singularity scaling exponent of a multiplicative cascade
β	binary number identifier
b	branching number
b_s	spatial branching number
b_t	temporal branching number
C_α	set of α-elements
$c(\alpha)$	codimension function of measures with singularity scaling exponent of order α
Cov	covariance operator
d	embedding dimension of the Euclidean space
dx	probability measure which is generated by the FM approach on the x axis
$dx\|_\theta$	rotated probability measure resulting from applying a transformation to the measure dx with a rotation angle θ
dy	probability measure which is generated by the FM approach on the y axis
$dy\|_\theta$	rotated probability measure resulting from applying a transformation to the measure dy with a rotation angle θ
Δ	increment

Δ_n^i	i-th subdomain after n partitions of a random cascade
ε_n	discrete random variable
$\mathbb{E}$	expectation value
ε	rate of kinetic energy dissipation
$f(\alpha)$	fractal dimension of measures with singularity scaling exponent of order α
$g_\lambda(\mathbf{x}, t)$	green function/scaling transformation group
Γ_λ	random cascade generator at the scale λ
G	set identifier
h	power spectrum exponent
H	self-similarity scaling exponent/anisotropic scaling exponent
H_x	zonal anisotropic scaling exponent
H_y	meridional anisotropic scaling exponent
I^k	numeric interval of order k
λ	spatial or timescale
$K(q)$	moment scaling exponent function for the moments of order q
ℓ	spatial longitude
L_0	length of the spatial domain
L_n	length of the spatial domain
λ_0	integral scale
S_q	structure function of order q
k	wave number
η	Kolmogorov scale
μ	measures (normalized) of rainfall
μ_∞	limit mass (or measure) of a random cascade
$\mu(\Delta_n)$	low frequency component of a random cascade outcome
M_μ^q	moment scaling function of order q for the measures μ
M	cumulative measure function
N	number of elements in a set/number of contractile affine maps
$g_\mu(\mu)$	probability density function of measures μ
ξ	power spectrum exponent
$\xi(q)$	scaling exponent function for the structure function of order q for the rainfall field
θ_q	scaling exponent function for the structure function of order q for the rainfall field
$\zeta(q)$	scaling exponent function for the structure function of order q for the velocity field
q	order of the statistical moment
$\mathbb{P}_r$	probability
P	power spectrum
R	Rainfall field
R_0	initial rainfall intensity in a random cascade
$\mathfrak{R}$	real numbers
$\mathfrak{R}^+$	real positive numbers
θ	rotation angle for the multifractal measures projection
τ	time scale/temporal lag
$\tau(q)$	mass exponent scaling function for the moment q
t	continuous time position
T	transpose operator
x	spatial position for the standard basis e_1

$\mathbf{x}$	spatial vector $\mathbf{x} = \sum_i^d x_i e_i$
$\mathbf{u}$	velocity vector $\mathbf{u} = \sum_i^d u_i e_i$
V	random variable
W	random variable/random cascade generator
W_n	contractile map
y	spatial position for the standard basis e_2
Z_∞	high-frequency component of a random cascade outcome

Abbreviations

FM	fractal multifractal
pdf	probability density function
SLLN	strong law of large numbers

Author details

Victor M. Peñaranda-Vélez[1*†], Carlos E. Puente[2†], Óscar J. Mesa[3†] and Arturo I. Quintanar[1†]

1 Instituto de Ciencias de la Atmósfera y Cambio Climático - UNAM, México City, México

2 University of California, Davis, USA

3 Universidad Nacional de Colombia, Medellín (Antioquia), Colombia

*Address all correspondence to: victor.penaranda@atmosfera.unam.mx

†These authors contributed equally.

References

[1] Fractals FJ. New York and London. London: Plenum Press; 1988. p. 283. DOI: 10.1007/978-1-4899-2124-6

[2] Falconer K. Fractal Geometry: Mathematical Foundations and Applications. 2nd ed. Chichester, West Sussex: Wiley; 2003. p. 337

[3] Frisch U, Parisi G. On the singularity structure of fully developed turbulence. In: Ghil M, Benzi R, Parisi G, editors. Turbulence and predictability in geophysical fluid dynamics and climate dynamics. Amsterdam, New York: North-Holland Publishing Company; 1985. pp. 84-88. ISBN: 978-0444869364

[4] Frisch U. Turbulence, the Legacy of a. N. Kolmogorov. New York: Cambridge University Press; 1995. p. 312

[5] Meneveau C, Sreenivasan K. Simple multifractal cascade model for fully developed turbulence. Physical Review Letters. 1987;**59**(13):1424-1427. DOI: 10.1103/PhysRevLett.59.1424

[6] Meneveau C, Sreenivasan KR. The multifractal nature of turbulent energy dissipation. Journal of Fluid Mechanics. 1991;**224**:429-484. DOI: 10.1017/S0022112091001830

[7] Kolmogorov AN. Local structure of turbulence in an incompressible liquid for very large Reynolds number. Proceedings of the Royal Society A. 1991; **434**:9-13. DOI: 10.1098/rspa.1991.0075

[8] Corrsin S. On the spectrum of isotropic temperature fluctuations in an isotropic turbulence. Journal of Applied Physics. 1951;**22**:469-473. DOI: 10.1063/1.1699986

[9] Harris D, Foufoula-Georgiou E, Droegemeier KK, Levit JJ. Multiscale statistical properties of a high-resolution precipitation forecast. Journal of Hydrometeorology. 2001;**2**:406-418. DOI: 10.1175/1525-7541(2001)002<0406:MSPOAH>2.0.CO;2

[10] Menabde M, Harris D, Seed A, Austin G, Stow D. Multiscaling properties of rainfall and bounded random cascades. Journal of Water Resources Research. 1997;**33**(12): 2823-2830. DOI: 10.1029/97WR02006

[11] Tessier Y, Lovejoy S, Schertzer D. Universal multifractals: Theory and observations for rain and clouds. Journal of Applied Meteorology and Climatology. 1993;**32**:223-250. DOI: 10.1175/1520-0450(1993)032<0223:UMTAOF>2.0.CO;2

[12] Verrier S, Mallet C, Barthès L. Multiscaling properties of rain in the time domain, taking into account rain support biases. Journal of Geophysical Research: Atmospheres. 2011;**116**: D20119-1-D20119-16. DOI: 10.1029/2011JD015719

[13] Foufoula-Georgiou E. On scaling theories of space-time rainfall: some recent results and open problems. In: Barndorff-Nielsen OE, Gupta VK, Pérez-Abreu V, Waymire E, editors. Stochastic Methods in Hydrology: Rain, Landforms and Floods. Advanced Series on Statistical Science and Applied Probability. Vol. 7. Singapore: World Scientific Publishing Co. Pte. Ltd; 1998. pp. 25-72. ISBN: 978-9810233679

[14] Lovejoy S, Schertzer D. Multifractals and rain. In: Kundzewicz ZW, editors. New Uncertainty Concepts in Hydrology and Water Resources. Cambridge: Cambridge University Press; 1995. pp. 61-103. DOI: 10.1017/CBO9780511564482. Availbale from: http://www.

physics.mcgill.ca/~gang/eprints/eprint Lovejoy/neweprint/Rainall1995.pdf [Accessed: 01 March 2023]

[15] Lovejoy S, Schertzer D. Multifractals, universality classes and satellite and radar measurements of cloud and rain fields. Journal of Geophysical Research: Atmospheres. 1990;**95**(D3):2021-2034. DOI: 10.1029/JD095iD03p02021

[16] Over T, Gupta V. Statistical analysis of mesoscale rainfall: Dependence of a random cascade generator on large–scale forcing. Journal of Applied Meteorology and Climatology. 1994;**33**:1526-1542. DOI: 10.1175/1520-0450(1994)033 <1526:SAOMRD>2.0.CO;2

[17] Schertzer D, Lovejoy S. Physical modeling and analysis of rain and clouds by anisotropic scaling multiplicative processes. Journal of Geophysical Research: Atmospheres. 1987;**92**(D8): 9693-9714. DOI: 10.1029/JD092iD08p09693

[18] Tessier Y, Lovejoy S, Hubert P, Schertzer D, Pecknold S. Multifractal analysis and modeling of rainfall and river flows and scaling, causal transfer functions. Journal of Geophysical Research: Atmospheres. 1996;**101**(D21): 26427-26440. DOI: 10.1029/96JD01799

[19] Veneziano D, Furcolo P. Multifractality of rainfall and scaling of intensity–duration–frequency curves. Water Resources Research. 2002;**38**(12): 42-1-42-12. DOI: 10.1029/2001WR000372

[20] Evertsz CJG, Mandelbrot BB. Multifractal measures (Appendix B). In: Peitgen H, Jürgens H, Saupe D, editors. Chaos and Fractals: New Frontiers in Science. New York: Springer-Verlag; 1992. pp. 921–954. DOI: 10.1007/978-1-4757-4740-9

[21] Veneziano D, Langousis A, Furcolo P. Multifractality and rainfall extremes: A review. Water Resources Research. 2006; **42**(6):W06D15-1-W06D15-18. DOI: 10.1029/2005WR004716

[22] Hubert P, Tessier Y, Lovejoy S, Schertzer D, Schmitt F, Ladoy P, et al. Multifractals and extreme rainfall events. Geophysical Research Letters. 1993;**20**(10):931-934. DOI: 10.1029/93GL01245

[23] Cortis A, Puente CE, Sivakumar B. Encoding hydrologic information via a fractal geometric approach and its extensions. Stoch Environ Res Risk Assess. 2010;**24**(5):625-632. DOI: 10.1007/s00477-009-0349-4

[24] Gupta V, Waymire E. A statistical analysis of mesoscale rainfall as a random cascade. Journal of American Meteorological Society. 1993;**32**:251-267. DOI: 10.1175/1520-0450(1993)032 <0251:ASAOMR>2.0.CO;2

[25] Huang H, Puente CE, Cortis A. Geometric harnessing of precipitation records: Reexamining four storms from Iowa City. Stoch Environ Res Risk Assess. 2013;**27**(4):955-968. DOI: 10.1007/s00477-012-0617-6

[26] Kang B, Ramrez JA. A coupled stochastic space–time intermittent random cascade model for rainfall downscaling. Water Resources Research. 2010;**46**(10):W10534-1-W10534-17. DOI: 10.1029/2008WR007692

[27] Lovejoy S, Schertzer D. The Weather and Climate: Emergent Laws and Multifractal Cascades. New York: Cambridge University Press; 2013. p. 475

[28] Marzan D, Schertzer D, Lovejoy S. Causal space-time multifractal processes: Predictability and forecasting of rain fields. Journal of Geophysical Research:

Atmospheres. 1996;**101**(D21):333-346. DOI: 10.1029/96JD01840

[29] Maskey ML, Puente CE, Sivakumar B. Deterministic simulation of highly intermittent hydrologic time series. Stochastic Environmental Research and Risk Assessment. 2017; **31**(10):2719-2732. DOI: 10.1007/s00477-016-1343-2

[30] Over T, Gupta V. A space-time theory of mesoscale rainfall using random cascades. Journal of Geophysical Research: Atmospheres. 1996;**101**(D21): 319-331. DOI: 10.1029/96JD02033

[31] Peñaranda-Velez VM. Representación geométrica determinística de registros de precipitación puntual con el modelo Fractal – Multifractal [thesis]. Bogotá: Universidad Nacional de Colombia; 2008

[32] Puente CE. A new approach to hydrologic modeling: Derived distribution revisited. Journal of Hydrology. 1995;**187**:65-80. DOI: 10.1016/S0022-1694(97)87978-4

[33] Puente CE, Obregon N. A deterministic geometric representation of temporal rainfall: Results for a storm in Boston. Water Resources Research. 1996;**32**:2825-2839. DOI: 10.1029/96WR01466

[34] Puente CE, Sivakumar B. Modeling geophysical complexity: A case for geometric determinism. Hydrology and Earth System Sciences. 2007;**11**(2): 721-724. DOI: 10.5194/hess-11-721-2007

[35] Paschalis A, Molnar P, Fatichi S, Burlando P. A stochastic model for high–resolution space–time precipitation simulation. Water Resources Research. 2013;**49**(12):8400-8417. DOI: 10.1002/2013WR014437

[36] Cortis A, Puente CE, Huang H, Maskey ML, Sivakumar B, Obregón N. A physical interpretation of the deterministic fractal–multifractal method as a realization of a generalized multiplicative cascade. Stoch Environ Res Risk Assess. 2014;**28**(6):1421-1429. DOI: 10.1007/s00477-013-0822-y

[37] Barnsley M. Fractals Everywhere. 2nd ed. San Diego: Academic Press Inc.; 1988. p. 394

[38] Deidda R. Rainfall downscaling in a space-time multifractal framework. Journal of Water Resources Research. 2000;**36**(7):1779-1794. DOI: 10.1029/2000WR900038

Chapter 10

Fractal and Multifractal Characterization of 3D Video Signals

Amela Zekovic

Abstract

Video signals are responsible for the largest amount of information in data storage and data transmission systems. Three-dimensional (3D) video formats are the largest in terms of the amount of data, and thus the required bit rate. For the efficient transmission of 3D video in communication systems, a detailed knowledge of the traffic characteristics of the format is necessary. In this research, characterization of 3D video signals is performed by using fractal and multifractal analyses. Codes for analyses are written in MATLAB and Python. Communication network traffic shows self-similar behavior with long-range dependence (LRD). Using visual methods and a rigorous statistical methods, it was shown that video sequences of 3D video formats have fractal self-similar properties with LRD and high Hurst parameter values. It is shown that investigated 3D video formats have multifractal structure by using the histogram method. The research included different 3D video formats, specifically with one video sequence (frame compatible, FC and frame sequential, FS) and with two or more video sequences (multiview, MV). Cases with different signal qualities defined by the quantization parameter, different types of frames, different groups of picture (GoP), and different broadcasting methods were taken into account and separately analyzed.

Keywords: 3D video, video transport, transport models, Hurst index, multifractal analysis, inverse multifractal analysis

1. Introduction

Three-dimensional (3D) video contains multiple views of the scene, which provides a viewer with a sense of depth [1–4]. 3D video formats with one video sequence are frame compatible (FC) and frame sequential (FS) formats, while formats with multiple video sequences are marked as multiview (MV) formats [5–7].

One of the main features of 3D video content is a very large amount of data, which is large compared to other content and also compared to conventional single-view video. Such a large amount of data represents a limitation in the possibilities of storing and transmitting 3D video in communication systems such as cellular networks, digital television, or all IP networks. This is why it is especially important to know the detailed characterization of 3D video formats for its transmission.

3D video transmission research includes testing statistical properties, required protocols, quality of service, transmission modeling, and adaptation to different communication networks. Previous research is often devoted to the analysis of protocols for delivering 3D video representation formats [8] or on video quality as a central topic [9]. Appropriate traffic models are often created for more efficient video transmission [10]. The latest research includes using machine learning for adaptation and more successful transport of 3D video [11].

In this chapter, the results of the research of 3D video format with two views in MV video format, FS format, and FC side-by-side (SBS) representational format are given. Publicly available, long video traces with frame data [12, 13] were used. One video sequence of frames contained 51,200 frames in full high definition (HD). In addition to examining the 3D video format, the research included the determination of characteristics for 3D video with different streaming methods.

One of the directions of image and video research follows the presence of fractal features in these signals [14]. A particularly significant fractal feature that is present in images and videos is self-similarity, which has found application in the processing of these signals, for example, in fractal compression [15, 16]. In Refs. [17, 18], it was shown that traffic in communication networks has fractal behavior.

This book chapter analyzes the fractal self-similarity of 3D video representation formats using graphical and a more rigorous statistical method. The fractal self-similar nature of the 3D video format is demonstrated using the Hurst parameter. The obtained values show that 3D video sequences exhibit long-range dependent (LRD) behavior during transmission [19].

The application of fractal properties to the network traffic characterization offers simple models but also limits their accuracy. As the traffic with variable bit rate leads to the creation of a structure with changes on different scales, a more detailed characterization of the traffic is possible using multifractals [20, 21]. In this research, the characterization of the 3D video format was also performed using multifractal parameters. Multifractal characterization of 3D video signals is given in Refs. [22–25]. It is shown that the most pronounced multifractal nature characterizes the MV video, while it is the least pronounced for the FS video. A detailed multifractal analysis of the 3D video format included an examination of the influence of the frame type, organization of the frames in group of picture (GoP), and a broadcasting method on the multifractal properties.

This book chapter is divided into seven sections. After the introductory remarks, Section 2 provides an overview of 3D video representation formats. Section 3 provides the basics of video traces, their structure, and use, as well as data on the used video traces of 3D video signals. In Section 4, the fundamentals of complex systems are described, and details about fractals and multifractals are given. The results of the fractal self-similar analysis of the 3D video formats, with explanations of the methods used, are given in Section 5. The multifractal characterization of the 3D video formats and the results of this analysis are given in Section 6. Final remarks with special emphasis on the contributions of this chapter are given in Section 7.

2. 3D video formats

A three-dimensional (3D) video is a video that creates a sense of depth. This is one of the latest video formats, which presence in systems ranges from initial applications for cinema and television to modern ones for mobile users and transmission through IP networks.

The 3D video transmission system includes acquisition and processing, 3D video representation and coding on the transmitter side, transmission through the system, then decoding, view generation — rendering, and display on the receiver side [26–28]. A representation format of 3D video largely determines the modules in the system.

In order to ensure successful transmission through different communication systems and enable their compatibility, international standards are needed. Organizations ITU-VCEG (International Telecommunication Union-Video Coding Experts Group) and ISO-MPEG (International Organization for Standardization-Moving Picture Experts Group) published a large number of necessary standards in the field of digital media, including standards for 3D video.

2.1 Classification of 3D video formats

3D video formats can be classified based on the number of video sequences into video formats with one video sequence and 3D video formats with two or more video sequences.

2.1.1 3D video formats with one video sequence

3D video formats with one video sequence include frame-compatible (FC) 3D video and frame-sequential (FS) 3D video.

Traditional video encoders can be used for 3D video formats with one video sequence, such as the H.264/MPEG-4 standard adopted by the joint body joint video team (JVT) formed by ITU-T and ISO/MPEG. The H.264/MPEG-4 standard (H.264 or MPEG-4 Part 10 standard) is also referred to as H.264/AVC (H.264/MPEG-4 Advanced Video Coding).

FC 3D video formats are formats with two views of the scene (stereo video formats) where the frames from the individual views are multiplexed into a single encoded frame. Within this format, half of the samples correspond to the left view of the 3D video, while the other half corresponds to the right view, meaning that each view has half of the full frame resolution [1].

There are several different packings of left and right view samples into a single frame, such as splitting the resolution into two parts (horizontal resolution splitting for side-by-side (SBS) format and vertical for top-bottom (TB) format), interleaving of left and right views (column and row interleaved formats) and packing according to the principle of a chessboard (chessboard format).

Enhancement of FC video parameters can be achieved during coding by adding supplemental enhancement information (SEI) within H.264/AVC or using scalable extension within the code scalable video coding (SVC), which represents extension of the H.264/AVC standard.

FS video formats are created by encoding individual views of 3D video according to the principle used to encode classic video with a single video sequence in full resolution. The frames obtained in this way are combined with each other forming one video sequence [1]. This offers better quality not only in the case of the FC format but also a larger amount of the data.

2.1.2 3D video formats with multiple video sequences

3D video formats with multiple video sequences include multiview (MV) 3D video format, multiview video plus depth (MVPD), and layered video plus depth (LVPD).

A video containing two or more views of the same scene, where each view is a single video sequence, is MV 3D video. Unlike the FC 3D video format, where each of the frames has full high definition (HD) [5, 29], more video sequences within a video lead to a significant increase in the amount of data, that is. the required bandwidth for transmission. In order to reduce the amount of data of the MV 3D video, during its coding, in addition to temporal and spatial redundancy within one view of the scene, the redundancy that largely exists between views is also used. The concept with interview prediction was included in the H.264/AVC standard as an amendment multiview video coding (MVC) [5]. Within this coding, intra-coded (I) frames, predictive (P) frames, and bi-directive (B) frames are used for reference frames, as shown in **Figure 1**.

3D video formats that, in addition to video sequences, contain additional depth data are MVPD video formats. Additional depth data is placed on the map based on the distance value of a point on the scene in the image plane to the camera. Using the depth matrix makes it much easier to create artificial views of the scene, but creating precise depth values is very complex. The advantage of depth video is direct compatibility with existing 2D systems.

Video with multiple video sequences has a significantly larger amount of data, so to improve this feature LVPD format was created, where image data from different views and depth data from different views placed in layers with the aim of reducing redundancy. Image information is reduced by projecting images from different views onto each other and eliminating repetitive content [30–32]. This procedure is repeated for depth data as well. In addition to the main layers, additional data about hidden elements of the scene are created in the image domain and in the depth domain.

Support for 3D video formats with depth is provided by the standard MPEG-C Part 3. This format requires the synthesis of additional views of the scene on the receiver side.

3. Video traffic using video traces

3.1 Video traffic characterization

There are generally three approaches to analyze video traffic in communication networks: using video bit stream, using video traces, and using a video transport model.

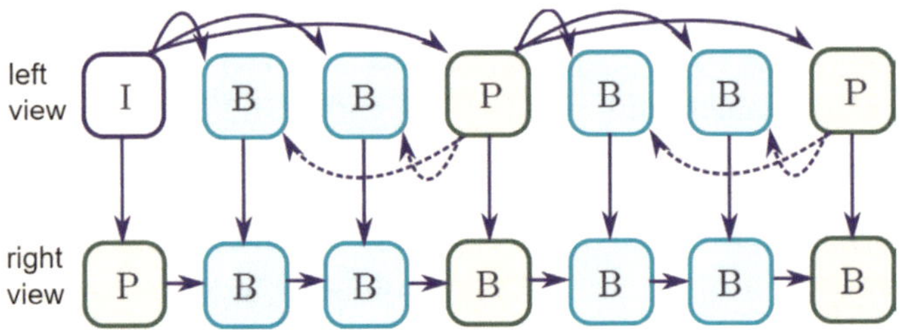

Figure 1.
Illustration of interview prediction between views in MV 3D video.

After encoding, a video bit stream is obtained with the selected encoder parameters. This data can be used for subjective and objective evaluations of video quality, as well as for video traffic analysis. During these analyses, the documents used are of the order GB, which greatly complicates the examination. Also, most of the video material is proprietary, which limits research.

Based on the coded video material, it is possible to create a video trace, where the specific bits in the video bit stream are not saved but the information at the frame level. This approach does not violate the rights of the video owner.

The characterization of video sequences includes the determination of statistical characteristics of video signal traffic at the frame level and at the level of a group of images, as well as quality characteristics [13, 33]. These parameters are determined based on traffic metrics, quality metrics, as well as by rate-distortion (RD) and variability distortion (VD) curves [33].

3.2 Creating video traces

The creation of video traces begins by feeding the video signal to the reference encoder, which parameters can be changed. As a result of encoding, an encoded bit stream and a video trace document are obtained. Video trace documents are most often used to test the effectiveness of the coding method and to characterize the traffic of video material [33].

To generate YUV uncompressed video content from the stored video content, conversion software can be used, such as FFmpeg. A video trace can also be generated on the basis of pre-encoded video content present on the Internet, for example, using the software tool MPlayer.

Video traces for long video sequences, suitable for traffic evaluation, are available online thanks to researchers. Good examples are the database of the Technical University of Berlin published by Fitzek and Reisslein (http://www-tkn.ee.tu-berlin.de/research/trace/trace.html) and the video database Arizona State University traces credited by Reisslein, Karam, Seeling, Fitzek, and Madsen (http://trace.eas.asu.edu).

3.3 Structure of video traces

Video traces contain data about encoded video content. There are two types of video traces: video traces for frame sizes and video traces for frame quality.

A video trace document is typically a text document with ASCII (American Standard Code for Information Interchange) or unicode characters. At the beginning of the document, there is a header that gives general data about the video, and below is the video trace data in tabular form [33].

The video trace document header usually contains video title, frame format—spatial resolution, and frame rate, encoding method and encoder used, group of picture (GoP) organization, quantization parameter values, and basic statistical data.

Each row of the tabular part of the trace refers to one frame in the video. The first column contains the frame numbers $n = 0, \ldots, N - 1$, the second the display time of the frame t_n in cumulative form, $t_n = nT$, and the type of frame (I, P, or B) in the third column. The size of the frames is described by the parameter X_n^q, which carries information about the number of bytes in the nth frame that is compressed with the quantization parameter q. In the video trace, for each of the frames, the quality is expressed within the parameters $Q_n^{q,Y}$, for the luminance component, and $Q_n^{q,U}$ $Q_n^{q,V}$, for chrominance components.

In addition to the above data, the video trace can also contain additional data such as offset distortion and scores for the video quality metric, as well as the ordinal number of views to which the frame belongs in 3D video.

3.4 Parameters of analyzed 3D video sequences

An overview of the parameters of the examined 3D video formats is given in **Table 1**. MV, FS, and FC SBS 3D video formats were investigated using long, publicly available, video traces encoded using JMVC version 8.3.1, that is. H.264 reference software JSVM version 9.19.10 in single-layer encoding mode. The tested 3D video had two views ($V = 2$), a left view (LV) and a right view (RV), where each view has 51,200 frames in full HD resolution (1920×1080) and frame rate $f = 24$ frames/s. Tim Burton's film Alice in Wonderland, which is a combination of real and computer-generated film, was used to evaluate the performance of 3D video. The analysis included different encoder settings for quantization parameters: $q_p = (I, P, B)$: (24,24,24), (28,28,28), and (34,34,34). The length of a group of picture (GoP) for MV and SBS format is 16 frames, while for FS format it is 32 frames, which allows equal display time between adjacent I frames for each format. The picture group scheme is B1, which means that one B frame is located between consecutive I and P frames. Additionally, for MV 3D video GoP G16B7 was analyzed. The test also included determining the characteristics of video sequences for individual types of frames.

In addition to examining the 3D video formats, the research included the determination of characteristics for 3D video with different streaming methods. Frames can be transmitted in their original form one by one or view-by-view for MV video when the LV and RV are viewed individually as separate video sequences. The second streaming method implies that certain views are combined, sequentially (S) or with aggregation (combining, C). In the case of sequential merging, frames from individual views are used to create a single sequence with the following frame order: frame 1 from view 1, frame 1 from view 2, …, frame 1 from view V, frame 2 from view 1, frame 2 from view 2, … In the case of merging with aggregation, MV frames are formed, where one MV frame represents the sum of all frames with the same serial number from different views. This principle was applied at the level of aggregation with two frames (marked as MV-C 2 and FS-C 2) and at the level of aggregation with the length of a group of images (MV-C 16 and FS-C 16).

Parameter	LV	RV	MV	SBS	FC
Frame rate f [frames/s]	24	24	48	24	48
Frame format	HD 1920×1080				
Number of frames	51,184	51,184	102,368	51,184	102,368
Video standard	H.264/AVC MVC			H.264/AVC	
Encoder	JMVC 8.3.1			JSVM 9.19.10	
Quantization parameters $q_p(I, P, B)$	(24,24,24), (28,28,28), (34,34,34)				
Group of Pictures, GoP	G16B1, G16B7				G32B1

Table 1.
Parameters of analyzed 3D video sequences.

4. Complex system theory

4.1 Fundamentals of complex systems

A complex system represents a network of individual components without central control, where simple individual components and their actions lead to complex behavior of the system as a whole, to sophisticated information processing, and possibilities for development and potentially adaptation.

Examples of complex systems include many natural and made systems such as insect colonies, neural networks, immune systems, cities, economic markets, and communication networks [34, 35].

One of the basic geometric properties of these systems is great complexity and fine structure regardless of the scale of observation [14]. By using fractal geometry, it is possible to measure the complexity of the system [36].

4.2 Fractals and fractal dimensions

Fractals represent complex geometric shapes, which have a fine structure on an arbitrarily small scale. Fractals usually have a certain degree of self-similarity, which can be exact or statistical [14, 37].

In order to measure the degree of complexity, it is necessary to find a relationship between how quickly the length, area, or volume change with decreasing scale. This would further allow determining the required size depending on the scale. The law that gives this relationship is a power law in the form

$$y \sim x^D. \tag{1}$$

This law is the basis of the fractal self-similarity dimension, which is defined as:

$$D = \frac{\log m}{\log r}, \tag{2}$$

where m is the number of copies, and r is the scaling factor.

The principle of measuring complexity using the fractal self-similar dimension is simple to calculate, but for the case, where fractals do not scale, a generalized definition of the fractal dimension—box dimensions is needed.

Calculating the box dimension involves placing a complex structure on the grid whose boxes are ε in size and counting the boxes that contain part of the structure $N(\varepsilon)$. By reducing the size of the ε box, a correspondingly larger value for $N(\varepsilon)$ is determined. By plotting the obtained values on a log/log diagram and determining the slope of the line passing through the measured values, the box dimension is obtained.

In addition to the listed dimensions, which are used the most, there are other dimensions such as the pointwise dimension, a correlation dimension, and the Hausdorff dimension [14, 38].

4.3 Multifractals and multifractal spectra

In some situations, it is necessary to associate a certain measure with part of the set. Such as fractal sets, measure sets are considered to be highly irregular and variable in their value at different scales. If the variability of these sets is exactly or statistically self-similar, we are talking about multifractals [39].

To characterize the complexity of multifractals is necessary to include weighting coefficients that will depend on the value of the measure μ.

The size α defined as

$$\alpha = \frac{\log \mu(\text{box})}{\log(\varepsilon)} \tag{3}$$

represents the coarse Hölder exponent. The parameter α gives the ratio of the logarithm of the measure within the box to the logarithm of the size of the box. Typically, the coarse Hölder's exponent α has values in the range $[\alpha_{min}, \alpha_{max}]$, where $0 < \alpha_{min} < \alpha_{max} < \infty$.

Once the parameter α is determined, it is necessary to determine its frequency distribution. For different values of the parameter α, the number of boxes $N_\varepsilon(\alpha)$ that have a rough Hölder exponent equal to α is determined. As in the determination of features of the original set, and with the parameter α the probability distribution with decreasing box size $\varepsilon \to 0$ has no limiting value. Therefore, a weighted logarithmic function is needed.

$$f_\varepsilon(\alpha) = -\frac{\log N_\varepsilon(\alpha)}{\log(\varepsilon)}. \tag{4}$$

This function for $\varepsilon \to 0$ tends to the function $f(\alpha)$.

The function $f(\alpha)$ describes how with the decrease in the size of the boxes ε, the number of boxes with a coarse Hölder exponent equal to α, $N_\varepsilon(\alpha)$ The connection of these parameters is given by the relation $N_\varepsilon(\alpha) \sim \varepsilon^{-f(\alpha)}$. The function $f(\alpha)$ describes the distribution of the parameter α. The graph $f(\alpha)$ is denoted as multifractal spectrum or $f(\alpha)$ curve.

Sometimes the parameter α is denoted as the strength of the singularity and then $f(\alpha)$ represents the singularity or the Hausdorff singularity, and the $f(\alpha)$ curve is denoted as the spectrum of the singularity. The singularity α tracks local changes in the signal and $f(\alpha)$ gives the global characteristics of the data [39–41].

For an empirical self-similar measure, only one n-th state of the measure is known. In this case, in order to determine the multifractal spectrum, the previous states of the measurement are reconstructed by coarse-graining of the measurement. For discrete data, the smallest measure is obtained for the n-th measure state when the size of the boxes is $\varepsilon = 1$. The sum of all measures within a state equals one or is normalized to one.

To estimate multifractal characteristics and multifractal spectra of 3D video formats, the histogram method is used. This method is characterized by slower convergence of $f_\varepsilon(\alpha)$ to $f(\alpha)$ and slower execution but can enable the display of additive processes in the signal and enables inverse multifractal analysis—determining the exact pieces of data that have a selected pair of values $(\alpha, f(\alpha))$. The histogram method works all the time with pure original data and contains fewer approximations.

5. Self-similarity analysis of 3D video formats

In this research, the fractal self-similarity of 3D video formats was examined using visual and rigorous statistical methods [19].

5.1 Self-similarity parameters

A process with values $X(t), t \in \mathbb{R}$ is self-similar with parameter H if for each positive factor, c the distributions of finite dimensions are $X(ct), t \in \mathbb{R}$, equal to the distribution $c^H X(t), t \in \mathbb{R}$ [14, 17]. Thus, typical self-similar series or self-similar processes look qualitatively the same regardless of the time scale on which they are observed. This does not mean that the same image is repeated in an identical way, but it is a general impression that the process remains the same.

One of the advantages of the self-similar model is that the degree of self-similarity is expressed by only one number, the H parameter. This parameter is also referred to as the Hurst exponent or Hurst index. The values of the Hurst parameter can be divided into three categories [14, 42], where range $0 < H < 1/2$ stands for short-range dependent (SRD) processes, $H = 1/2$ for random walk, and range $1/2 < H < 1$ for long-range dependent (LRD) processes.

It is shown that the traffic in communication networks exhibits self-similar properties, by measuring self-similarity using the Hurst parameter. This parameter indicates the level of traffic explosiveness, where higher values of the parameter indicate greater traffic explosiveness [43].

5.2 Methods for testing self-similarity

Self-similarity testing methods include the aggregated variance method, the R/S statistic method, and the method based on the correlation of multifractal analysis and the Hurst parameter, that is. multiscale method and others [17, 43]. In this chapter, the results will be given using the R/S statistic method, which was previously successfully tested on a process with fractal Gaussian noise, with a predefined value of Hurst parameter.

The R/S statistic method involves the following steps. Data on frame sizes X_n, $n = 1, \dots, N$ are divided into K nonoverlapping blocks. Then, the rescaled adjusted ranges $R(t_i, d)/S(t_i, d)$ are determined for multiple values of the parameter d, where $t_i = N/K(i-1) + 1$ are the starting points of the blocks that fill condition $(t_i - 1) + d \leq N$,

$$R(t_i, d) = \max\{0, W(t_i, 1), \dots, W(t_i, d)\} - \min\{0, W(t_i, 1), \dots, W(t_i, d)\} \tag{5}$$

wherein,

$$W(t_i, k) = \sum_{j=1}^{k} X_{t_i + j - 1} - k\left(\frac{1}{d}\sum_{j=1}^{d} X_{t_i + j - 1}\right), \quad k = 1, \dots, d \tag{6}$$

and $S^2(t_i, d)$ the variance of the data $X_{t_i}, \dots, X_{t_i + d - 1}$. R/S values were determined for each value of the d parameter. This number decreases with the increase in the value of the d parameter, which is a consequence of the restrictions that exist for the t_i values. By drawing a graph of the value $\log(R(t_i, d)/S(t_i, d))$ as a function $\log d$, the so-called R/S diagram is obtained. The slope of the regression line of the R/S value provides an estimate of the Hurst parameter.

5.3 Results of 3D video self-similarity analysis

Frame size traces of MV 3D video are shown in **Figure 2(a)**, as a relation between the size of the frames and their sequence number. The video trace is also shown for

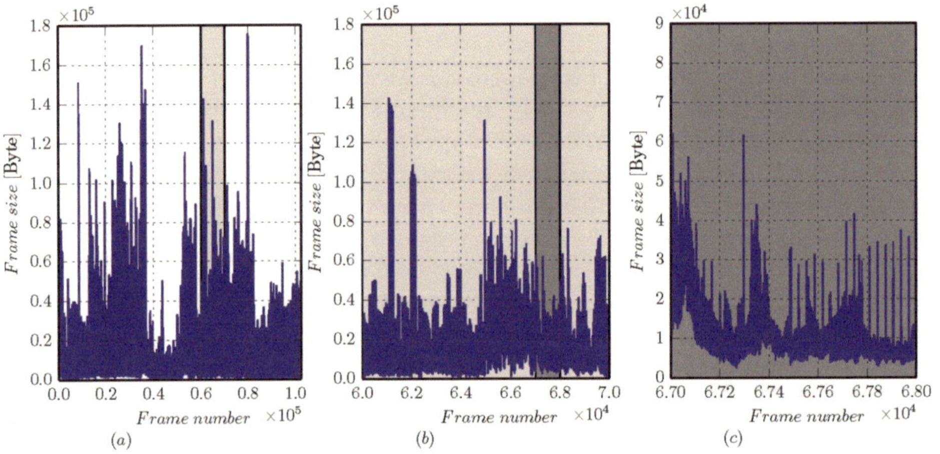

Figure 2.
Frame size in relation to frame number for MV 3D video: (a) full frame size video trace, (b) enlarged section of part of the trace labeled as (a), and (c) further enlarged section of part of the trace labeled as (b).

two smaller ranges of frame numbers in **Figure 2(b)** and **(c)**. The selected ranges are marked in **Figure 2**. Qualitatively, the self-similarity of MV 3D video is observed in **Figure 2**.

This self-similarity test is only graphical, so more rigorous statistical methods are needed to show the self-similar nature of 3D video formats, such as R/S statistic method.

Graphic representation of the R/S statistic method is shown in **Figure 3**. This is an example of a plot used in the estimation of the Hurst parameter. Reference lines have been added to the graphs for comparison, specifically reference lines with slopes $k_1 = 1/2$ and $k_2 = 1$. The slope of the regression line β was estimated using the method of least squares fit and the R/S statistical method $H = \beta$. Examples of H parameter estimation given in **Figure 3** refer to MV 3D video representation format with quantization parameters $q_p(I, P, B) = (28,28,28)$.

Complete Hurst index evaluation results for left view (LV), right view (RV), combined view (CV), multiview (MV) 3D video formats, side-by-side (SBS), and frame sequential (FS) format are given in **Table 2**. The values of the quantization parameters for all video formats are $q_p(I, P, B) = (28,28,28)$. Based on the results given in **Table 2**, it is concluded that Hurst index for 3D video formats has large values, indicating a high level of long-range dependence process for 3D video content.

The previous results apply to video content when frames are broadcast one by one. **Table 3** gives the results for the analysis of self-similar nature for streaming 3D video where aggregation of frames in pairs (denoted by C-2) or aggregation at the level of group of pictures (denoted by C-16). The results show a self-similar LRD behavior of 3D video content with aggregation, but with generally smaller values of the Hurst parameter.

6. Multifractal analysis of 3D video formats

Video content is characterized by high variability and the appearance of explosiveness, especially in the case when the values of the encoder's quantization

3D Video type	Video format	$H_{R/S}$
One video sequence	SBS	0.9601
One video sequence	FS	0.9881
Multiview video	LV	0.9627
Multiview video	RV	0.9774
Multiview video	CV	0.9737

Table 2.
Hurst index values for 3D video representation formats using the R/S statistic method.

3D Video type	$H_{R/S}$	3D Video type	$H_{R/S}$	3D Video type	$H_{R/S}$
SBS-C 2	0.9504	FS-C 2	0.9860	CV-C 2	0.9811
SBS-C 16	0.8676	FS-C 16	0.8567	CV-C 16	0.8681

Table 3.
Comparison of Hurst index values for 3D video representation formats with frame pair aggregation (C 2) and GoP aggregation (C 16).

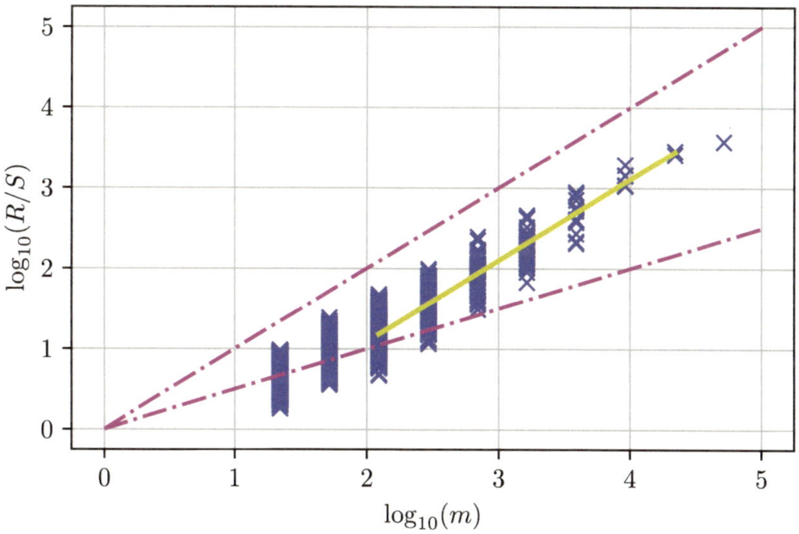

Figure 3.
Evaluation of Hurst parameter for MV 3D video using R/S statistic method.

parameters are constant. Characterization of traffic variability is possible with statistical methods, such as coefficient of variation, RD, and VD curves. The dynamics and tendencies of fractal structure changes can be described using self-similarity parameters, as was done for 3D video in Chapter 5.3.

The mentioned characterizations of 3D video formats are very significant, but for the complete characterization of 3D video, multifractal analysis is also necessary due to the complexity of the video process. Also, this method enables a more precise insight into the behavior of the signal and its variability when broadcasting frame-by-frame or view-by-view, and gives particularly important results for broadcasting with frame aggregation. Such a more precise characterization of the 3D video format can allow for better modeling of this type of traffic, as well as for improving the efficiency of video transmission methods.

Multifractal characterization of 3D video signals is given in Refs. [23–25].

6.1 Estimation of multifractal properties by histogram method

In this research, the histogram method was used for multifractal characterization of 3D video formats.

The histogram method for determining the multifractal spectrum starts with overlaying a self-similar measure with boxes of size ε. In the case of this method, $f_\varepsilon(\alpha)$ slowly tends to $f(\alpha)$, so for a better estimation of the spectrum, sliding boxes were used to cover the measure instead of nonoverlapping ones.

In this analysis, $n_\varepsilon = 8$ of different box sizes were used, with the following values $\varepsilon = [1,3,5,9,13,21,29,37]$, that is, indices for ε_k are $k = 1, 2, \dots, n_\varepsilon$. For each value of ε_k, the overall measure, $\mu_{i,k}$, where $i = 1, 2, \dots, n$ is determined. The total length of the data and the value for ε_k determine the parameter n, where a larger number of existing measures is obtained for smaller boxes. For more efficient computation, all measurements are placed in a matrix M, where the size of the matrix is given by $[n_\varepsilon \times n]$, for the largest possible n (the smallest value for ε_k). The coarse Hölder's exponent α_i, $i = 1, 2, \dots, n$ is now determined as the slope for the graph $\log(\mu_i, k)$ in relative to $\log(\varepsilon_k/L)$, where L is the length of one-dimensional data.

The range of α values, $[\alpha_{min}, \alpha_{max}]$ is discretized into $D = 100$ parts of equal length $\Delta\alpha$, and the α_d values are formed based on interval center values. In the domain of the parameter α of the value α_i, for different values j, $j = 1,3,5, \dots, 199$, the number $N_j(\alpha_d)$ is determined as the number of boxes of size j, which have the value α_i in the region $\Delta\alpha$ around α_d. This procedure is repeated for all values of α_d, $d = 1, 2, \dots, 100$. Finally, the $f(\alpha_d)$ values are determined as the slopes of the graph $-\log(j/n)$ versus $\log(N_j(\alpha_d))$. A plot of $f(\alpha_d)$ versus α_d represents an estimate of the $f(\alpha)$ curve.

To understand the discussion of multifracral properties, it is important to know that α_{min} corresponds to the largest values of the measure, while α_{max} is associated with the smallest and smoothest part of the signal under investigation.

6.2 Results of multifractal analysis of 3D video format

3D video representation formats were examined in a multifractal sense using the histogram method. Multifractal video spectra are determined considering different multiview video views, different 3D video formats, different video streaming approaches, frame types, and GoP organization.

Multifractal spectra determined using the histogram method for different views of multiview 3D video and for different 3D representation formats are given in **Figure 4(a)** and **(b)**, respectively. Multiview 3D video views have different data complexity (sequences of frame sizes), where RV has the lowest complexity, while CV and LV have similar, higher complexity, as seen by the range of dimensions in the spectrum. Also, the maximum of the multifractal spectrum for RV is the largest (the case with the highest frequency), while the singularity $\alpha(f_{max})$ is the smallest compared to LV and CV. Multifractal spectra for CV multiview 3D video, FS, and SBS 3D representation formats show that the widest spectrum (most complex structure) occurs for CV, followed by SBS, while the least complexity is found in FS format.

The multifractal spectrum for RV has two dominant peaks at the top of the spectrum, which is due to the two processes that exist within the data— the P and B frames in the signal formed using the frames from the LV as a reference. A similar, but less pronounced process is present in the case of the spectrum for CV of MV 3D video.

Multifractal spectra obtained by applying the histogram method for selected types of frames: Only I frames, only P frames, and only B frames for different 3D video representation formats are given in **Figure 5**. The smallest changes in the multifractal spectrum are observed for I frames as a consequence of the similar coding principle of this type of frame, regardless of the 3D video format. The maximum of the multifractal spectrum is the smallest in the case of I frames, followed by the maximum for B-type frames and P-type frames. The smallest decrease of the multifractal spectrum around its maximum was observed for I frames, which is a consequence of the structure with a larger number of dimensions whose participation in the structure is significant. Multifractal spectra for P and B frames have a faster decay and fewer dimensions with a large participation in the signal. For isolated frame types, sequences with only P frames have a multifractal spectrum with the smallest value of the α_{min} parameter, which means the highest explosiveness. Similar conclusions can be drawn based on the multifractal spectra obtained using the moment method. The relationships between multifractal spectra for CV, SBS, and FS formats for I-only, P-only, and B frames are the same as for video sequences with all frame types.

Comparison of multifractal spectra of MV 3D video with different groups of picture (GoP) organization is given in **Figure 6**. GoP with higher number of B frames shows slightly less additive processes in signal, but width of multifractal spectra stays approximately the same.

As given in Ref. [13] based on CoV, the FS 3D format has better properties in terms of variability than the MV 3D format, but with frame aggregation (for a pair of consecutive frames or for 16 consecutive frames—one GoP) video sequences show better characteristics in terms of variability for both formats, especially for MV 3D video which becomes more similar in characteristics to the FS format. These results

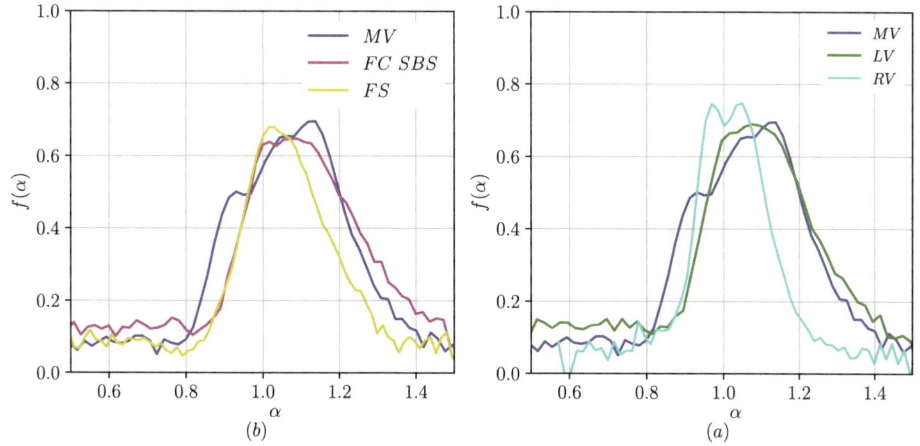

Figure 4.
Multifractal spectra obtained using the histogram method: (a) different representational formats of 3D video and (b) different views for MV 3D video — Left, right, and combined.

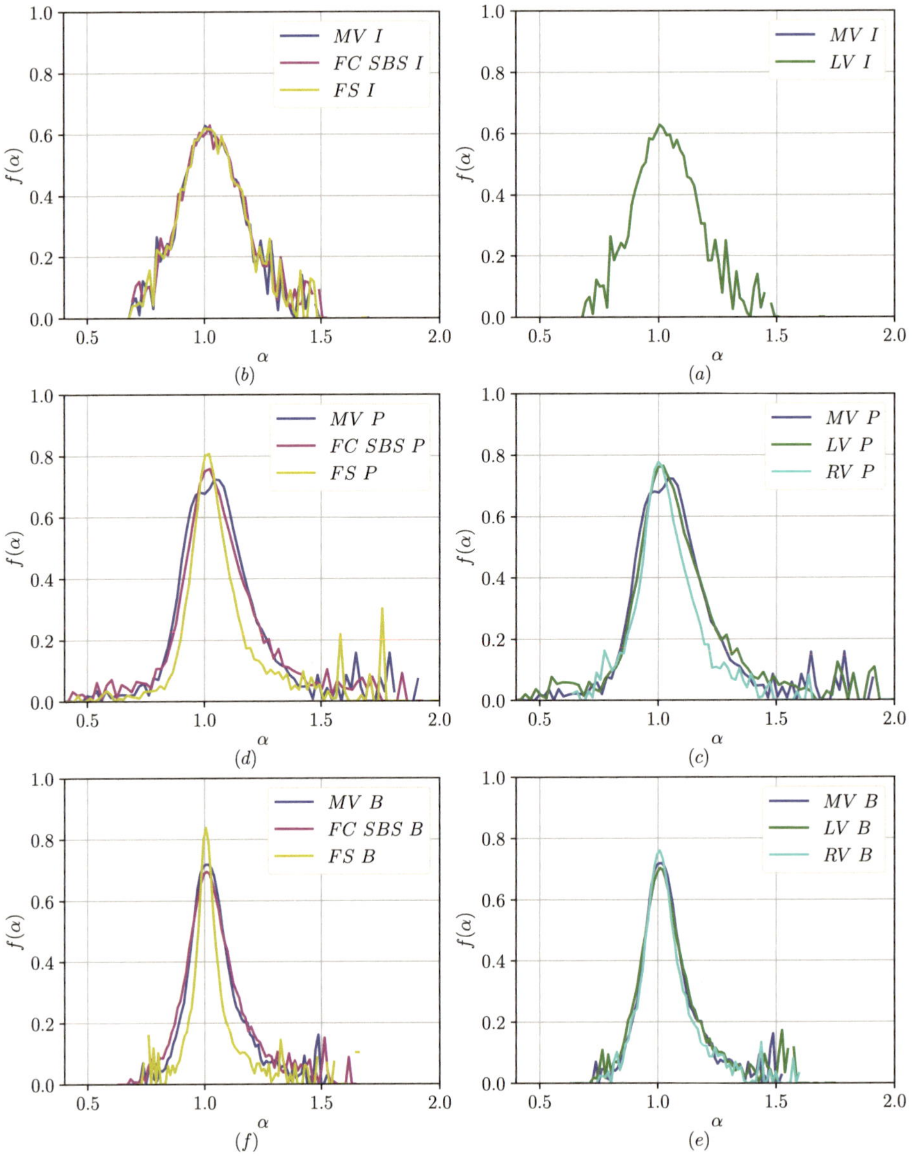

Figure 5.
Multifractal spectra obtained by the histogram method for selected types of frames: (a) only I frames for 3D video representation formats, (b) only I frames for different views of MV 3D video, (c) only P frames for 3D video representation formats, (d) only P frames for different views of MV 3D video, (e) only B frames for 3D video representation formats, and (f) only B frames for different views of MV 3D video.

were further analyzed in terms of the explosiveness of the video signal, which is very important for data traffic. Video sequences with frame aggregation were also analyzed using the multifractal spectrum obtained using the histogram method. These results are shown in **Figure 7**. The main change on the spectrum, for video content with aggregation, is the position of the spectrum maximum $\alpha(f_{max})$, which represents the complexity of the structure in the case with the highest frequency. Both 3D formats,

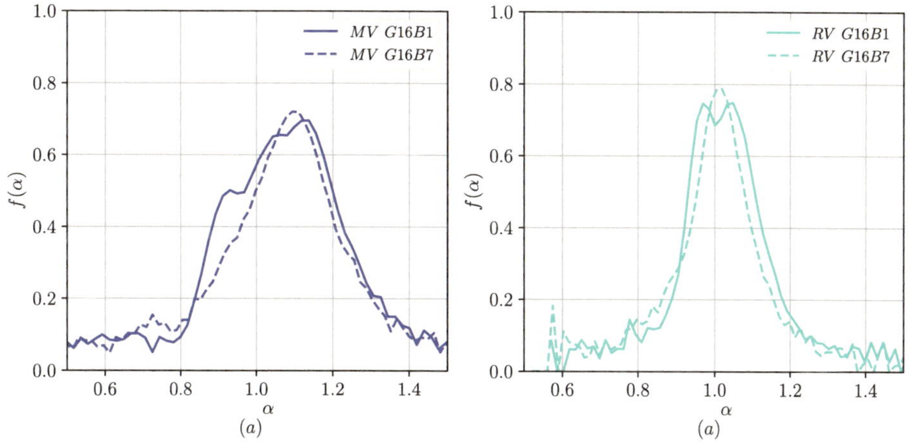

Figure 6.
Multifractal spectra by the histogram method: (a) MV 3D with GoP G16B1 and GoP G16B7, (b) RV 3D with GoP G16B1 and GoP G16B7.

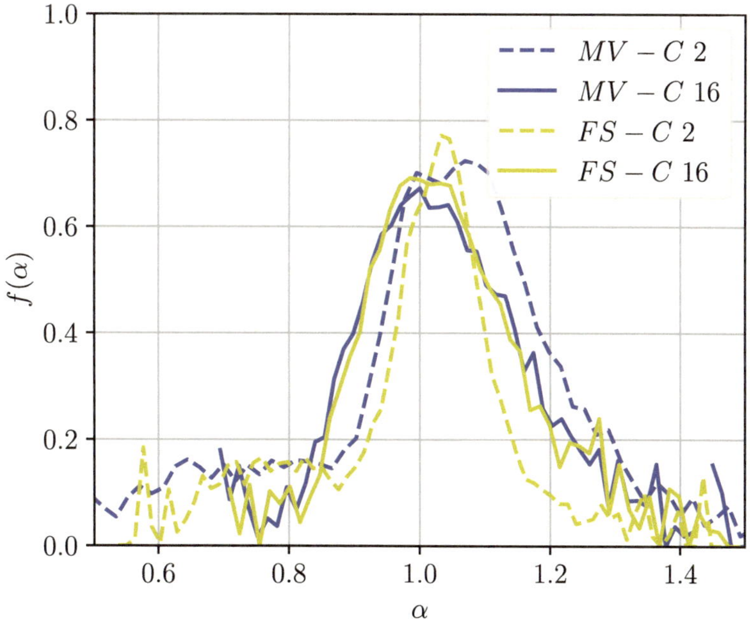

Figure 7.
Multifractal spectra by the histogram method for MV and FS video with aggregation.

FS and CV, have a less complex structure, a smaller value of α in the case with the highest frequency, for sequences with aggregation. A small improvement was observed for aggregation at the level of two frames, while a larger improvement was observed for aggregation at the new group of images. The parameter f_{max} for FS and CV spectra is very similar.

7. Conclusions

In this book chapter, 3D video representation formats are characterized: multiview (MV) video representation format that uses multiview video coding, as well as frame sequential (FS) and frame compatible (FC) and side-by-side (SBS) formats that use encoding using a conventional single-view video encoder. The analysis of 3D video formats is realized using 3D video traces for these three main 3D video representation formats. Publicly available traces of long video sequences [12, 13] were used. Video traces contained information about each individual video frame: display time, frame size, frame type, and frame quality parameters. Within one 3D video sequence for one setting of quantization parameters, there were two views with 51,200 frames each in full HD 1920 × 1080 (resolution)

In this book chapter, several aspects regarding fractal and multifractal characterization of 3D video are presented.

An essential overview of 3D video representation and compression formats according to fractal and multifractal properties is given [19, 25].

It has been shown that video sequences of 3D video format have fractal self-similar properties with long-range dependence (LRD) [19]. The characterization of the fractal properties of 3D video was realized using a visual method and detailed calculation using R/S statistic method using codes implemented in MATLAB and Python, which showed high values of the Hurst parameter of about 0.95 indicating a high-level explosiveness and LRD characteristics [19].

It is shown that the three main examined 3D video formats have a multifractal structure [23–25], as well as individual views in multiview 3D video format using the multifractal spectra. The implementation of the algorithm and the application of the histogram method using codes written in the MATLAB and Python [25] showed that MV video has the most pronounced multifractal structure for the most pronounced singularity (the largest part of the structure), while the least pronounced multifractal nature is present in FS video.

An overview of the fractal and multifractal properties of the 3D video format depending on the choice of frame type was given, which showed a great similarity of the multifractal properties of the intra-coded (I) frames of the 3D video format, which is a consequence of the similar coding principle of this type of frame, and which showed differences in the variability of frame types, especially pronounced for prediction (P) coded and bidirectional (B) coded frames for different 3D video formats, which is a consequence of different coding approaches, as well as the highest explosiveness of P-type frames [22, 23, 25].

Fractal and multifractal properties of 3D video representation formats in the case of frame aggregation and in different conditions of 3D video broadcasting were exposed with the aim of determining more efficient transmission, which showed a decrease in the expression of variability if only certain generalized dimensions or fractal self-similarity are observed, but an increase in explosiveness on a smaller to the part of the signal that is observed thanks to the multifractal spectrum [19, 25].

Values of fractal and multifractal parameters of 3D video representation formats could be used for modeling 3D video traffic in communication systems [17, 19, 20, 25, 37, 39].

The results presented in this chapter can be used to improve methods of reducing variability (smoothing) and to design more efficient multiplexing methods, in applications such as smoothing with pre-sending part of the signal [44], more precisely in terms of explosive traffic estimation and its management process and controls.

DOI: http://dx.doi.org/10.5772/intechopen.1002377

Statistical multiplexing methods that use frame type information to improve features can potentially improve performance given the multifractal features of the different frame types presented in this book chapter [22, 23, 25]. Also, the results can be used in examining the impact of the introduction of 3D video formats into the multiplexer with 2D video formats on the channel characteristics.

Author details

Amela Zekovic
Department School of Electrical and Computer Engineering, Academy of Technical and Art Applied Studies Belgrade, Belgrade, Serbia

*Address all correspondence to: amelaz@viser.edu.rs

References

[1] Merkle P, Müller K, Wiegand T. 3D video: Acquisition, coding, and display. IEEE Transactions on Consumer Electronics. 2010;**56**(2):946-950

[2] Rocha L, Gonçalves L. An overview of three-dimensional videos: 3D content creation, 3D representation and visualization. In: Bhatti A, editor. Current Advancements in Stereo Vision. Rijeka: IntechOpen; 2012

[3] Eisert P, Schreer O, Feldmann I, Hellge C, Hilsmann A. Volumetric video – acquisition, interaction, streaming and rendering. In: Valenzise G, Alain M, Zerman E, Ozcinar C, editors. Immersive Video Technologies. Cambridge, Massachusetts: Academic Press; 2022. p. 289-326

[4] Kerdranvat M, Chen Y, Jullian R, Galpin F, François E. The video codec landscape in 2020. In: The future of video and immersive media. ICT Discoveries. International Telecommunication Union. 2020;**3**(1): 73-84

[5] Tech G, Chen Y, Müller K, Ohm JR, Vetro A, Wang YK. Overview of the multiview and 3D extensions of high efficiency video coding. IEEE Transactions on Circuits and Systems for Video Technology. 2016;**26**(1):35-49

[6] Karsten Müller PE, Schwarz H, Wiegand T. Video Data Processing. Heidelberg, Germany: Springer-Verlag - part of Springer Nature; 2019

[7] Punchihewa A, Bailey D. A review of emerging video codecs: Challenges and opportunities. In: 2020 35th International Conference on Image and Vision Computing New Zealand (IVCNZ). Wellington, New Zealand: IEEE. 2020. pp. 1-6

[8] Mohib H, Swash MR, Sadka AH. Multi-view video delivery over wireless networks using HTTP. In: Proceedings of International Conference on Communications, Signal Processing, and their Applications. Sharjah, United Arab Emirates: IEEE; 2013. pp. 1-5

[9] Malekmohamadi H. Automatic subjective quality estimation of 3D stereoscopic videos: NR-RR approach. In: 3DTV Conference: Capture, Transmission and Display of 3D Video. Copenhagen, Denmark: IEEE; 2017. pp. 1-4

[10] Tanwir S, Nayak D, Perros H. Modeling 3D video traffic using a Markov modulated gamma process. In: 2016 International Conference on Computing, Networking and Communications (ICNC). Kauai, HI, USA: IEEE; 2016. pp. 1-6

[11] Hao J, Liu S, Guo B, Ding Y, Yu Z. Context-adaptive online reinforcement learning for multi-view video summarization on Mobile devices. In: 2022 IEEE 28th International Conference on Parallel and Distributed Systems (ICPADS). Nanjing, China: IEEE; 2023. pp. 411-418

[12] Video Trace Library. Available from: http://trace.eas.asu.edu.

[13] Pulipaka A, Seeling P, Reisslein M, Karam L. Traffic and statistical multiplexing characterization of 3D video representation formats. IEEE Transactions on Broadcasting. 2013;**59** (2):382-389

[14] Peitgen H, Jürgens H, Saupe D. Chaos and Fractals. New York: Springer; 1992

[15] Ingole AV, Kamble SD, Thakur NV, Samdurkar AS. A review on fractal

compression and motion estimation techniques. In: 2018 International Conference on Research in Intelligent and Computing in Engineering (RICE). San Salvador, El Salvador: IEEE; 2018. pp. 1-6

[16] Tiwari M, Chaurasia V, Siddiqui EA, Patel V, Kumar A, Patankar M. Enhanced image compression using fractals and principle component analysis. In: 2023 1st International Conference on Innovations in High Speed Communication and Signal Processing (IHCSP). Bhopal, India: IEEE; 2023. pp. 502-507

[17] Sheluhin O, Smolskiy S, Osin A. Self-Similar Processes in Telecommunications. New York: John Wiley & Sons, Inc.; 2007

[18] Filaretov GF, Chervova AA. Fractal characteristics of the network traffic. In: 2022 VI International Conference on Information Technologies in Engineering Education (Inforino). Moscow, Russian Federation: IEEE; 2022. pp. 1-4

[19] Zekovic A, Reljin I. Self-similar nature of 3D video formats. In: Jonsson M, Vinel AV, Bellalta B, Belyaev E, editors. Proceedings of Multiple Access Communications - 7th International Workshop, MACOM 2014, Halmstad, Sweden, August 27–28, 2014. Lecture Notes in Computer Science series. Vol. 8715. New York City, USA: Springer; 2014. pp. 102-111

[20] Dang TD, Molnár S, Maricza I. Capturing the complete multifractal characteristics of network traffic. In: Global Telecommunications Conference. Taipei, Taiwan: GLOBECOM IEEE; 2002. pp. 2355-2359

[21] Millán G, Osorio-Comparán R, Lomas-Barrie V, Lefranc G. The associative multifractal process: A new model for computer network traffic flows. In: IEEE Chilean Conference on Electrical, Electronics Engineering, Information and Communication Technologies. Valparaíso, Chile: IEEE; 2021. pp. 1-6

[22] Zekovic A, Reljin I. Multifractal and inverse multifractal analysis of multiview 3D video. In: Proceedings of 21st Telecommunication Forum (TELFOR 2013), 26–28 November 2013; Belgrade, Serbia: TELFOR; 2013. pp. 753-756

[23] Zekovic A, Reljin I. Inverse multifractal analysis of different frame types of multiview 3D video. TELFOR Journal. 2014;**6**(2):121-125

[24] Zekovic A, Reljin I. Comparative analysis of multifractal properties of H.264 and multiview video. In: Proceedings of 1st International Conference on Electrical, Electronic and Computing Engineering IcETRAN 2014, 2–5 June 2014; Vrnjačka Banja, Serbia: IcETRAN; 2014. pp. EKI1.7.1-4

[25] Zekovic A, Reljin I. Multifractal analysis of 3D video representation formats. EURASIP Journal on Wireless Communications and Networking. 2014; **2014**(181):1-14

[26] Smolic A. 3D video and free viewpoint video – From capture to display. Pattern Recognition. 2011;**44** (9):1958-1968

[27] Zhang C, Su M, Liu Q, Yang M. 3D communication system integrating 3D reconstruction and rendering display. In: 2022 5th International Conference on Advanced Electronic Materials, Computers and Software Engineering (AEMCSE). Wuhan. China: IEEE; 2022. pp. 167-170

[28] Patil VTM, Anami B, Billur D. Effective 3D video streaming using 2D video encoding system. In: 2022 2nd Asian Conference on Innovation in Technology (ASIANCON). Ravet. India: IEEE; 2022. pp. 1-5

[29] Müller K, Merkle P, Tech G, Wiegand T. 3D video formats and coding methods. In: Proceedings of the 17th IEEE International Conference on Image Processing (ICIP): 26–29 September 2010; Hong Kong, China: IEEE; 2010. pp. 2389-2392

[30] Cagnazzo M, Pesquet-Popescu B, Dufaux F. 3D video representation and formats. In: Dufaux F, Pesquet-Popescu B, Cagnazzo M, editors. Emerging Technologies for 3D Video: Creation, Coding, Transmission and Rendering. New Jersey, USA: Wiley Publishing; 2013

[31] Yilmaz GN, Battisti F. Depth perception prediction of 3D video for ensuring advanced multimedia services. In: 2018 - 3DTV-Conference: Capture, Transmission and Display of 3D Video. Helsinki, Finland: IEEE; 2018. pp. 1-3

[32] Merkle P, Müller K, Marpe D, Wiegand T. Depth intra coding for 3D video based on geometric primitives. IEEE Transactions on Circuits and Systems for Video Technology. 2016;**26** (3):570-582

[33] Seeling P, Fitzek FHP, Reisslein M. Video Traces for Network Performance Evaluation. Dordrecht: Springer; 2007

[34] Mitchell M. Complexity: A Guided Tour. Oxford, New York: Oxford University Press; 2009

[35] Easley D, Kleinberg J. Networks, Crowds, and Markets: Reasoning about a Highly Connected World. Cambridge, United Kingdom: Cambridge University Press; 2010

[36] Lloyd S. Measures of complexity: A non-exhaustive list. IEEE Control Systems Magazine. 2001;**21**(4):7-8

[37] Feder J. Fractals. New York: Springer Science – Business Media; 1988

[38] Strogatz SH. Nonlinear Dynamics and Chaos: With Applications to Physics, Biology, Chemistry, and Engineering. Cambridge, Massachusetts: Westview Press; 2001

[39] Evertsz A, Mandelbrot B. Multifractal Measures. In: Peitgen H, Jürgens H, Andrews P, editors. Chaos and Fractals. New York: Springer; 1992. pp. 849-881

[40] Véhel JL, Tricot C. On Various Multifractal Spectra. Progress in Probability. 2004;**57**:23-42

[41] Chhabra A, Meneveau C, Jensen V, Sreenivasan K. Direct determination of the f(α) singularity spectrum and its application to fully developed turbulence. Physical Review A. 1989;**40** (9):5284-5294

[42] Mandelbrot BB, van Ness JW. Fractional Brownian motions, fractional noises and applications. SIAM Review. 1968;**10**:422-437

[43] Leland W, Taqqu M, Willinger W, Wilson DV. On the self-similar nature of Ethernet traffic (extended version). IEEE/ACM Transactions on Networking. 1994;**2**(1):1-15

[44] Devi UC, Kalle RK, Kalyanaraman S. Multi-tiered, Burstiness-aware bandwidth estimation and scheduling for VBR video flows. IEEE Transactions on Network and Service Management. 2013;**10**(1):29-42